A Grain of Truth

A Grain of Truth

The Media, the Public, and Biotechnology

Susanna Hornig Priest

ROWMAN & LITTLEFIELD PUBLISHERS, INC.
Lanham • Boulder • New York • Oxford

ROWMAN & LITTLEFIELD PUBLISHERS, INC.

Published in the United States of America
by Rowman & Littlefield Publishers, Inc.
4720 Boston Way, Lanham, Maryland 20706
http://www.rowmanlittlefield.com

12 Hid's Copse Road
Cumnor Hill, Oxford OX2 9JJ, England

British Library Cataloguing in Publication Information Available

Library of Congress Cataloging-in-Publication Data

Priest, Susanna Hornig.
A grain of truth : the media, the public, and biotechnology / Susanna Hornig Priest.
p. cm.
Includes bibliographical references and index.
ISBN 0-7425-0947-8 (alk. cloth) — ISBN 0-7425-0948-6 (pbk. : alk. paper)
1. Biotechnology—Public opinion. I. Title

TP248.23 .P75 2001
660.6—dc21

00-059056

Printed in the United States of America

∞™ The paper used in this publication meets the minimum requirements of American National Standard for Information Sciences—Permanence of Paper for Printed Library Materials, ANSI/NISO Z39.48-1992.

Contents

Preface

Sometimes referred to as the "new genetics," biotechnology is revolutionizing both agriculture and medicine. Some of biotechnology's most dramatically envisioned applications cross the boundary between agriculture and medicine—as when a domesticated farm animal such as a sheep is genetically engineered to produce human medicine in its milk and then cloned for mass production—not to mention boundaries among biological species. This book looks primarily at the agricultural side of this revolution and is concerned with how information about these technologies interacts with our social structure (especially the structure of our mass media) and our cultural psychology in ways that cause some issues to be highlighted and others to fade from view, thereby influencing the public agenda.

Biotechnology means many things, and in some ways it is not too useful to lump them together. Our recent invention of and then the popularization of this new term suggest that even while public opinion makes distinctions between these various biological technologies and their myriad applications, we do think of modern biotechnologies as occupying the same (or at least closely contiguous) conceptual space. While the term can be extended to any technological manipulation of a biological process, ranging from wine-making to the construction of mechanical prostheses, it is now most closely associated with the active manipulation of the "genetic blueprint" provided by DNA, most commonly in the form of what has come to be called genetic engineering, or the artificial and purposeful combining of DNA from different species in novel ways. But it is not limited to such processes; clearly, cloning techniques—which work with the DNA from a single species and even, remarkably enough, a single individual—are also biotechnology.

In writing this book, I have envisioned many audiences. I write primarily for people who want to understand the relationship between mass media messages, cultural and social conditions, and the formation of public opinion, using the public's response to agricultural biotechnology in the United States as my central case. I hope that this book will be useful to students of the relationship between mass media and society, scientists and science students who want to understand why there is such a public outcry about their work, sociology and social psychology students who want to understand public opinion formation, and science journalists and journalism students who want to write more effectively about contemporary issues involving both science and society. It may also be useful to business and public relations specialists who want to do a better job managing the interface between their worlds and that of the public. And because I see biotechnology as raising many issues of modern democratic practice, I hope that some people who are simply interested in the challenges of the modern world will pick up this book.

I am certain some readers will see me as opposed to biotechnology, a modern-day secular humanist Luddite hiding behind a cloak of academic respectability. Others will see me as an apologetic lackey of industry, offering insights and strategies designed to help corporate America repress the masses. I have met members of each of these groups many times, and both kinds of encounters have been interesting and instructive. In truth I am neither opposed to, nor in favor of, increased use of biotechnology in agriculture. I find the proposition that people should be divided into one "camp" or another simplistic, and the available opinion research confirms this. But because beliefs about biotechnology are in many ways simply a function of other kinds of beliefs about government, science, religion, corporate interests, and environmental and social priorities, it seems reasonable to explain my own perspective briefly so readers will know whose work they are reading.

Like most Americans, I am comfortable with biotech in some applications, less so in others. I also share with them a deep faith in democracy, tempered by concerns for our future in terms of both environmental and social issues. I seem to see more of my fellow citizens engaged in environmental than in social causes, and so I am somewhat more concerned at this point for the latter than the former. I do not believe that current trends in the social distribution of resources and benefits are likely to be sustainable. I am suspicious of the motives of large corporations, but I do not see them as composed of evil people. And I am realistic about the limited ability of vast, politicized government bureaucracies to be flexible enough to respond to new challenges, like that which biotechnology represents, but I do not see them as composed of incompetents. Religion does not have much influence on me, but I also worry whether a vague humanism will provide enough moral vision to see us into the next century and be able to balance the profiteering of private companies and the intellectual pedestrianism of government agencies, and to overcome the social and economic inequities that threaten our democratic way of life.

Unlike many American intellectuals, however, I do not see science as the solution. I do not believe that science can provide values or settle political and social disputes. I do not believe that science and technology are always and necessarily benign, nor that social injustice will disappear if scientific knowledge is equally distributed. I do not believe the advancement of science and technology is equivalent to social progress, and I am cautious about accepting the proposition that our prosperity (economic and in other forms) is solely dependent on expanding the scientific sector. As a scholar and a person, I certainly do not believe that science and technology are independent of social forces or that science literacy consists of the absorption of a sufficient number of facts. But I do, in a very literal sense, strongly "believe" in science. Here in Texas, I have had a number of religious creationist students whose presence in my classrooms has, ironically enough, forced me to clarify in my own mind the significant distinction between what kinds of axioms they are willing to accept on faith and what kind I accept myself. I fully acknowledge that science, like religion, is a belief system. Even so, it is *my* belief system.

I do have a handful of relatives some distance away who still farm; if they have been the least bit helped by this particular scientific revolution, I have not heard of it.

Having confessed all this so that readers will not have to go through the pages of this book wondering what sort of thinly disguised ideological agenda lurks between its lines, let me emphasize that I write here primarily as a social scientist. I have no doubt that my values crop up in these pages from time to time, as they inevitably do for all scientists. Most of what I have to say is grounded in empirical evidence of one sort or another, whether collected by me or by others, whether qualitative observations or quantitative measurements. Since all social scientific methods have their limits, since the time I have had available to write this book has not been infinite, since available methodologies do not permit direct inference of cause-and-effect relationships involving media and public opinion, and since the data on this complex sequence of events will never be complete, interpretation that may sometimes seem to go beyond the available data is required. But the interpretations presented are intended to be social scientific in character; that is, I intend my arguments to be analytically and empirically descriptive of our actual social situation rather than simply speculations or statements designed to serve any particular activist agenda. I do, however, sometimes compare the current controversies in my mind with the idealized vision of democratic debate that I absorbed somewhere in elementary school.

The fact that I felt compelled to offer all these caveats might itself be taken as evidence of the deep interaction between the "new genetics" and an entire constellation of social, political, religious, and philosophical beliefs.

Many people and several institutions made my work on this set of issues, which extends back nearly a decade, both possible and much better than it otherwise would be. These include Paul B. Thompson, formerly at Texas A&M University but now at Purdue, who first called biotechnology to my attention as an issue worth scholarly consideration; Colin Allen and Gary Varner, whose collegial interaction with me at the Center for Science and Technology Policy and Ethics at Texas A&M was always intellectually stimulating; Fuller Bazer of the Institute for Biosciences and Technology, also affiliated with Texas A&M, whose financial support always came with complete and unqualified encouragement that we continue to explore all social and ethical issues associated with biotechnology as fully as possible; Robert Kennedy, then Vice President for Research at Texas A&M, who provided financial support for the Center through his office but more importantly consistently offered his personal encouragement that we should continue this work; Rachelle Hollander at the National Science Foundation, whose program supported myself and others in critical earlier phases of this research; Bruce Lewenstein at Cornell, with whom my collaborations on this subject have generally been informal but were always enormously helpful; and the Texas A&M University College of Liberal Arts. My errors and opinions are, of course, inevitably my own.

Thanks as well to my dozens of interested graduate and undergraduate students, both those who agreed with me and those who did not, without whose lively concern for these and related issues this book would certainly have been less valuable; and to my sons, Andy and Wes Hornig, who now as young men continue to give up portions of the time, good temper, and undivided attention that they might otherwise have had from me so that I could complete my work.

1

Introduction

Now, more than ever, with these technologies in their relative infancy, I think it's important that, as we encourage the development of these new food production systems, we cannot blindly embrace their benefits. We have to ensure public confidence in general, consumer confidence in particular, and assure farmers the knowledge that they will benefit.

—U.S. Secretary of Agriculture Dan Glickman,
remarks prepared for National Press Club address,
July 13, 1999 (Glickman 1999)

European resistance to the introduction of the products of biotechnology, especially genetically engineered foods, into their food supply has taken the United States by surprise, severely straining international trade relations in the process. Europeans have vigorously protested and in some cases banned the use of GMOs, or genetically modified organisms, including soybeans and other products previously exported in bulk from the United States (Soybeans, a relatively inexpensive and readily available source of high-quality protein and oil, are ubiquitous, found in dozens of other foods.) Because the way U.S. agriculture is organized now makes separating genetically engineered and nonengineered products difficult and expensive, U.S. producers have acted aggressively to protest this European resistance, which they claim is based on paranoia and misinformation.

Why is the European public so upset about genetically engineered products entering their food supply? Many observers blame scaremongering by the European press, yet extensive, well-funded, and very careful research has established that press accounts in Europe have been, if anything, more positive overall than press accounts in the United States (Gaskell, Bauer, and Durant 1999; Gutteling, Olofsson, Fjaestad, Kohring, Goerke, Bauer, and Rusan 1999). Is this reaction, then, a function of European ignorance? The same body of research has shown that those European countries where knowledge of biotechnology is highest tend to have the most negative press on these issues. A more reasonable question, posed recently by a pair of science journalists visiting the southern U.S. from France, is therefore this one: Why is the U.S. public so complacent on this issue?

While the effects of media messages on public opinion are easily overstated, we do know that the news calls our attention to some issues at the presumed expense of others (McCombs and Shaw 1972; Iyengar and Kinder 1987) and invites us to understand those visible issues in particular ways. We also know that people tend not to voice opinions when they feel they are in a minority (Noelle-Neumann 1993), and it is the news media that give us our sense of what is mainstream opinion and what is not. Sometimes they do this through reporting the results of opinion polls, but more often they communicate the boundaries of acceptable opinion simply by representing certain definitions and understandings as the "reality" of a given situation. A degree of inertia thus tends to be attached to mainstream media versions of events and priorities: the mediated reality is difficult to challenge.

Driven largely by the quasi-monopolistic Associated Press wire service, U.S. news tends to be fairly homogeneous across the nation, and it tends to represent the point of view of large institutions, including (in the case of agricultural biotechnology) the point of view of corporate and other institutional stakeholders (Priest and Talbert 1994; Lewenstein, Allaman, and Parthasarathy 1998). If there are reasonable objections to be raised to genetically engineered foods and other products of agricultural biotechnology, then this quite literally comes as news to many of us. Regardless of private reservations, public debate is unlikely to be stimulated without news media attention to the issues, and without public debate, public opinion is unlikely to coalesce around such a complex and intangible issue as the use of recombinant DNA technology in agriculture. The apparent complacency of the U.S. public with respect to biotechnology may be largely an illusion. We have not yet had enough of a debate for the outcome to be certain.

People in the United States do raise a broad range of questions about biotechnology when specifically asked to consider its acceptability, and this range is not particularly constrained to those issues raised by news accounts (Priest 1995). But in the normal course of events, it is unlikely that enough critical attention will be drawn to any given set of issues to stimulate public thinking without a certain level of media activity. This is doubly so for technically complex issues for which the average citizen has few, if any, alternative sources of information. Even the involvement of activist groups tends to be ineffective unless their activities generate substantial media attention, and then they must overcome the media's implicit presumption that their positions are (almost by definition) beyond the pale of rational thought. This is yet more difficult when the only sure route into the media is to engage in dramatic behavior, as the history of movements for social change in this country from Vietnam to civil rights to the environment amply demonstrates. And it is triply so when mainstream opinion can present science as being on its side.

This book is neither for nor against agricultural biotechnology, but it is pro-democracy. Democratic decision-making in this area will require substantive and ongoing public debate. In the long run, stakeholder institutions run an enormous risk by contributing (even inadvertently) to the suppression of this debate; backlash opinion could be severely negative when it finally does surface, as it almost inevitably will. This may already be happening, sparked by controversies erupting in other parts of the world. If biotechnology is in fact a benign technological force that will help solve food shortages, improve human nutrition, and fight disease, as many scientists working in this area clearly believe, then the sooner we have this debate, the better. If these new technologies do in fact entail some new challenges for environmental protection, biodiversity, species integrity, religious prohibitions, and human health, then we had best know these things up front.

Being "sold" agricultural biotechnology by sweeping such controversies under the rug is in no one's interests, least of all the longer-term interests of its promoters, both corporate and scientific. In this sense it is simply bad public relations. Unformed public opinion remains volatile. For science-related issues such as these, U.S. culture tends to be optimistic, and it is not surprising that the initial climate of opinion here seems to have leaned toward the positive rather than the negative. This climate is still likely to be unstable and unpredictable. Aside from this kind of strategic consideration, however, the lack of opportunity for informed public debate about these revolutionary technologies and their social and economic impact is a serious failure of democracy. It is a failure that makes our claim to being a free society with a free press somewhat hollow.

Exceptions to the hegemony of large-institution perspectives in the U.S. news about agricultural biotechnology are equally instructive. Objections to biotech on the basis of minority food preferences (for example, those vegetarians and conservative Jews who may not want to consume certain animal proteins unwittingly) are generally ignored. A proposal by the U.S. Department of Agriculture to allow genetically engineered foods to be labeled "organic" became controversial enough to be withdrawn because it was of direct concern to a type of consumer already much more actively concerned about food issues than the average American. Even so, it was not a controversy that became especially visible in the national press. On the other hand, the cloning of an agricultural animal (Dolly the sheep) by a Scottish scientist named Ian Wilmut became enormously controversial in part because of its apparent direct challenge to the sacredness of the biological individual in U.S. culture. And it received an unprecedented amount of media coverage.

Similarly, more recent publicity surrounding "terminator genes" inserted into genetically engineered crop plants to make them sterile, preventing saving of the seed from one year to the next, was highly constrained in the mass media in comparison to the Internet and alternative press discussions; nevertheless, the issue broke through in the national media in the form of a struggle that dramatically pitted large corporate interests against the autonomy of the individual farmer. In other words, biotech becomes news when it directly threatens what we hold sacred: the individual human being as an independent biological and economic unit. Less direct biological and economic threats, threats to non-mainstream values, and diffuse threats to the ecology of the planet appear less likely to result in public challenges. The sequence of events is also important; it is likely that both journalists and the public generally responded to the terminator controversy in ways conditioned by the publicity that surrounded Dolly. A new media and opinion frame was emerging that linked these events.

In general, how do the U.S. media treat biotechnology, and why? The answers to these questions provide insights into the dynamics that characterize the operation of U.S. news media much more generally, both for other stories involving science and technology and for news of other types. These dynamics are complex and occur at many levels, from individual to organizational to societal (Shoemaker and Reese 1996). For biotechnology, powerful institutional influences, the way that "objectivity" and scientific facticity are understood and presented in U.S. journalism, and the ambient cultural environment all play particularly important roles.

INSTITUTIONAL HEGEMONY

The news media in the United States are heavily weighted toward the point of view of large institutions. This is not the result of a plot. Rather, it is an indirect consequence of the First Amendment provision guaranteeing press freedom from governmental interference. Without governmental support, media are forced to compete in the marketplace for their very survival. For television programming, these are the dynamics that produce fare designed to appeal to the least common denominator. Print news works in exactly the same way. The objective is not excellence but audience share, because it is audience share that generates advertising revenue, and it is advertising revenue that keeps both the news and the entertainment media, print and broadcast, in business. Unlike residents of the U.K., we do not pay subscription fees for basic broadcast services. We do not pay the true cost of newspapers and magazines, and for Internet news we pay close to nothing. Advertising pays these costs.

Advertisers do not walk into newsrooms and tell journalists and editors what to do. However, newspapers and news magazines (just like broadcast news, sit-coms, talk shows, and now Internet services) are still powerfully constrained by the need to entice members of the public to be interested in their wares. It is to their economic benefit to appeal to their readers, and this supports an active concern about reporting on affairs of interest to those readers, including controversial ones. Serious investigative journalism in this country is rare to nonexistent, however, in part simply because it is expensive. Filling up news pages with material from readily available, inexpensive sources, including the wire service stories from the Associated Press, is much more practical. The same dynamics produce news weighted toward institutional sources who can afford to hire public relations and public information specialists, hold press conferences, and send out press releases (Gandy 1982).

Working journalists claim not to be influenced by press releases. Nevertheless, when a story must be filed on deadline, the perspectives on events represented by the ubiquitous large institutions that generate the bulk of these releases (corporations, government agencies, universities, police departments, even school systems) inevitably seeps in. And when a story must be propped up with quotes from willing spokespeople, these large institutions are the ones to be consulted. Other voices, generally those of individuals acting as individuals, are ignored or discounted. For agricultural biotechnology, these discounted voices may include those of individual scientists not speaking on behalf of a corporation or a corporate-funded university research program, as well as the voices of individual consumers and individual small farmers. Despite our nominal adherence to the democratic ideal, individuals acting as individuals have little opportunity to participate in media-based public debate. They can write letters to the editor or post material on the Internet, but these are clearly secondary influences.

Other voices, especially those most consistently critical of genetic engineering, are discounted, even when included. Their views may be reported to create the appearance of a "balanced" story, yet they are subtly positioned as representing fringe (if not lunatic) perspectives. In the case of science stories, critical positions may be represented as being not only non- but anti-scientific—and therefore unsupportable. They are included only to be delegitimized. The practice tends to lump all criticisms (and all critics) together. The more controversial and complex the story, the more the mainstream U.S. news media seem to retreat to the use of "safe" sources from larger, well-established institutions. Appearing to take sides on a controversy not only violates the objectivity norm of U.S. journalism, it might cost readers—and therefore advertising dollars.

Additional factors probably also contribute to the hegemonic domination of a single, large-institution perspective on the news. Increasingly few cities in the United States are served by more than one daily newspaper; increasingly few corporations own multiple news outlets (Bagdikian 1983; Alger 1998). Furthermore, most news organizations are themselves large institutions, with smaller stations and papers normally owned by large chains, so it is not necessary to postulate conspiracies to observe that they are likely to lean toward a large-institution perspective. Finally, journalists and editors are often less anxious to scoop important news than they are to be covering, out of the thousands upon thousands of events and issues they might cover on a given day, just about the same things that the competition is covering. This seems to reinforce their sense of their own good news judgement, as well as reduce the risk that a "maverick" story will prove to be a liability in some way—whether through being discredited or simply by being unpopular with readers or viewers. Some have seen the Internet as reversing this trend; others see it as the most hegemonic news institution yet, one operating on a global scale.

In short, the myth of the U.S. press as a vigorous watchdog and guardian of the public interest is a long way from the reality. In general, the motivation to investigate seems thwarted on a day-to-day basis by the demands of business as usual and the routinized practice of the profession of journalism (Tuchman 1978). Without it necessarily being newspeople's conscious intention, mainstream opinion becomes equated with the opinions of powerful interested parties, rather than either majority opinion, the conclusions of those who are best informed, or the views of those who have a legitimate special stake in the issue but may be less powerful. Under these circumstances, resistance to the status quo analysis breaks through into the public sphere only when fundamental values or beliefs are challenged or when especially dramatic events unfold. Biotechnology, however, has provided such circumstances on more than one occasion.

Of course, stakeholder views (whether on biotechnology or any other issue) deserve a prominent place in media accounts. But they do not necessarily deserve to displace all other perspectives. In the long term, these are the interests that have the most to lose should public opinion turn against them, whether on superficially rational grounds or not. Ironically, attempts to discredit the opposition can in many ways be self-defeating, especially when the stakeholder voices set the agenda and define the issues in ways that appear to sidestep rather than address public concerns. Many examples of this kind of "cross communication" between stakeholders and those they are trying to persuade can be found for biotechnology. Decades of research on persuasion have confirmed that messages that address both sides of a given controversy are more likely to be effective with most audiences in any event. In other words, even stakeholders with a persuasive intent should find it in their interests to see alternative viewpoints presented.

SCIENCE AND "OBJECTIVITY"

Many, if not most, of the issues surrounding agricultural biotechnology today are policy issues rather than strictly scientific issues. Many of them have to do with the ownership of genetic resources and whether private corporations will endeavor to manipulate the situation so that they have too much, even quasi-monopolistic, economic control over agriculture. This could be damaging to the viability of both the developed-world family farm and developing-world agriculture more generally. Whether grounded or groundless, these fears are at the root of much concern about agricultural biotechnology. Predicting these kinds of economic impacts is not done in laboratories, however, and deciding what to do about them is a question of politics and societal values, not science per se.

Other concerns have to do with possible environmental impacts: pest-resistant plants may also damage beneficial insects; herbicide-resistant crops may encourage farmers to increase their use of chemicals; increasing monoculture (planting of vast areas in one or a few nearly identical varieties) increases the risk of catastrophe. Genetically modified plants may cross with wild varieties with unpredictable effects. Some speculate that the modified DNA of these plants may not be as stable as the DNA of varieties bred in traditional ways. Since neither ecosystems nor the possible responses of the genome to artificially induced change are entirely understood, science cannot fully predict these outcomes. Still other questions involve more narrowly ethical considerations ranging from animal welfare concerns to dietary restrictions and even religious reservations (the notion that it is inherently wrong to "play God" by tinkering with natural DNA structures). The idea of cloning individual human beings has its own set of ethical problems, as does any suggestion of mixing human and nonhuman DNA.

No one has been able to rule out consequences for human health entirely. Some allergic individuals may no longer be certain which foods they can safely eat, for example. Some consequences could be much less direct. For example, additional stress put on dairy cows as a result of using hormone supplements produced through genetic engineering technology may raise their rate of infection; even if the milk they produce is by itself perfectly safe to consume, the rate of antibiotic use might increase, and it is believed that antibiotics fed to dairy animals can come through in the milk. This creates an increased risk that the antibiotics will eventually lose their effectiveness (for people, not just for the cows). But for the most part, it is the great range of economic, regulatory, environmental, and ethical issues, *rather than food safety or human health issues of the type science can eventually resolve,* that biotechnology's critics are raising.

Still, scientific authority has great rhetorical power in our society. One of the ways stakeholders try to maintain control over public opinion about biotechnology is by relying heavily on scientific data and science spokespeople in this discussion, which involves shifting the agenda to the more strictly scientific or safety aspects of the controversy. With these concerns highlighted, expert scientific opinion can be invoked on their side. Because scientific evidence is seen as "objective" fact, both cultural tradition and journalistic practice usually let it stand unchallenged, even where the scientists themselves may disagree. Just as a political candidate might try to set a campaign agenda of issues in areas of that candidate's strengths, biotech's promoters have tended to focus public rhetoric on those arguments they have the best chance of winning.

Of course, scientific facts (as well as the theories that explain them) are almost always in dispute; science advances both through the gradual accumulation of empirical evidence and through disputes over its interpretation that sometimes results in what the well-known historian of science Thomas Kuhn was the first to call "paradigm shifts." Among philosophers of science there continues to be active debate about the relationship between scientific data and the real natural world it is supposed to represent. The first scientific journals evolved out of the process of researchers' writing to the leaders of scientific societies to present their data and argue for their interpretations of that data. Even though, as a pragmatic matter, most scientific controversies are eventually resolved, the contemporary frontiers of research in any given field at any given moment are characterized—in fact, defined—by the existence of competing explanations.

Yet, in a world in which "scientific" is nevertheless equated with "factual," journalists cling to the habit of reporting any and all scientific results as truths. This can be very misleading. It is this habit that makes nutrition science, for example, seem quixotic. One day, butter is bad for us; the next day it is good. What is really going on in such cases within the research community is that the matter is never finally settled by any particular piece of research taken alone. Rather, an ongoing, iterative process eventually leads to a consensus among experts. But it is the individual research result that is most likely to be the subject of a large-institution press release, whether from a university, an independent research organization, or even a scientific journal. And it is the individual research result that seems most often to become the subject of a piece of science journalism. Within science, *objective* generally means supported by empirical evidence gathered in a systematic and unbiased way or by systematic theoretical analysis, rather than based strictly on investigator opinion. But this does not mean, of course, that there are no important differences of scientific opinion, especially in new areas of science and especially with respect to likely impacts on complex systems (such as ecological or socioeconomic ones). Biotechnology in particular presents a host of complex scientific and social scientific uncertainties. Journalism often does not know what to do about them (Friedman, Dunwoody, and Rogers, eds. 1999).

The journalistic tradition of objectivity is rather different from the scientific one and arose out of a reaction to the sensationalistic excesses of so-called yellow journalism. It originally represented a more or less direct (albeit incomplete) borrowing from scientific tradition (Nelkin 1995). What elevated journalistic objectivity from a style of writing to a professional norm was probably its adoption by the prestigious *New York Times* (Schudson 1978). However, economic forces were also at work. The development of the "penny press" in the nineteenth century, aimed at a truly mass audience and supported by advertising dollars, and the emergence of the wire news services around the period of the Civil War both created new economic pressures on publishers to maximize both their potential readership (in order in turn to maximize advertiser dollars) and the efficiencies of scale associated with electronic distribution of material. In journalism, objective now means both factual and politically neutral (or balanced). Facticity is understood as a given for science and neutrality as irrelevant. Scientific claims tend to be reported uncritically unless overt controversy has erupted.

The First Amendment was written to protect a partisan press that openly advocated particular political positions, but the mainstream press in the United States lost this character almost entirely as industrialization proceeded. Objectivity came to mean news devoid of political position—news for everybody. Today even the tabloids pretend to objectivity in this sense. Applied to science, this perspective seems to have blinded some journalists to the real character of both scientific objectivity and scientific consensus. The convergence—or perhaps we should say the clash—between the journalistic and scientific interpretations of objectivity comes into play for news about biotechnology in that it inclines journalists to report all scientific opinion as fact and ignore the controversial opinion altogether. Opinions within science that diverge are very problematic for science journalists. Either they must identify and report only the most mainstream scientific opinion, discounting all challenges, or they must follow the tradition of political reporting and identify opposing points of view, creating stories that provide equal balance between them. Neither approach does justice to the real complexity of scientific consensus.

How are journalists to report on controversies like cold fusion, where scientific consensus did not support the claims of individual scientists involved in highly publicized research at legitimate institutions, or global warming, where a robust scientific consensus does seem to support a certain interpretation of the data but some dissent remains? These would probably not be such impossible challenges were the journalists not beginning with the assumption that scientific facticity is much less fluid and problematic than in reality it is. So where some scientists think biotechnology is both promising and safe and others have reservations, journalists are at a loss. Equally or even more so, they tend to be at a loss where the real issues are policy—such as in questions of the acceptability of unknown risks—or ethics or economics, rather than science, and some of the key stakeholders are corporate interests rather than universities or research labs. One relatively safe way out of such dilemmas is to portray non-stakeholder opinions in such cases as representing unscientific positions while the larger-institution stakeholder positions are represented as objectively factual. This is safer in the sense that, by definition, non-stakeholders are less powerful and less able to effectively challenge such representations. It also fulfills the expectation that a single scientific explanation should exist.

None of this is to say that stakeholders are *necessarily* wrong about either biotechnology's promise for humanity or details of the associated science itself. Indeed, on average, we would probably expect stakeholders (including biotech researchers) to have more access to sound scientific evidence and to have thought more carefully about the issues. But, conversely, it should also be apparent that great danger lies in the assumption that stakeholder opinion is necessarily correct, or that scientific knowledge is the only appropriate foundation for policy in this area. In conjunction with the institutional biases noted above, the journalistic tradition of objectivity probably exacerbates a very worrisome tendency to see stakeholder opinion in this area as truth. And, again, this is probably not even in the stakeholders' own long term best interests. It is not the way to stimulate healthy public debate. Whatever promise biotech holds for solving human problems, without this debate it is unlikely to be fulfilled.

Journalistic accounts, even in the so-called elite press, are simply not at their most effective when dealing with science that is not yet established fact, or with the gray areas between science and policy, or with situations in which commercial interest and scientific truth are difficult to untangle. Agricultural biotechnology represents all three of these.

CULTURAL CONTEXT

All of these events take place within a broader cultural context that shades their interpretation, as well as influences their production. In U.S. culture, both agriculture and science are quasi-sacred endeavors. We were founded as an agrarian nation with a seemingly endless frontier and a firm belief that the autonomous individual frontiersman-farmer could meet and conquer all challenges the new world might present. Science and industry would help with this mission. American ingenuity and ambition—the willingness to develop and adopt new technologies of all kinds—would be what would put us ahead of our more tradition-bound European cousins. American democracy and the American economy would work because the advancing march of scientific research and technological development coupled with universal education would produce a nation of citizens capable of astounding levels of individual achievement. That emphasis on the individual deserves some attention here.

Our culture is intensely individualistic. Our achievements are seen as less the accomplishments of ourselves as a nation, or of our communities or families or schools, as they are of individuals acting independently. Genetic heritage is an important part of what defines us culturally as individuals; indeed, an argument can be made that our culture is obsessed with biological identity (Nelkin and Lindee 1995). While most news about biotechnology, especially agricultural biotechnology, remained nearly invisible on our national news agenda, it is by no means accidental that the issue of cloning exploded. Cloning did not emerge primarily as an agricultural issue or an issue of animal welfare, nor did it emerge straightforwardly as a technical development with important business ramifications for the pharmaceutical industry. This is true despite its having elements of all three of these. It emerged as a technique seen as foreshadowing the inevitable cloning of individual human beings, and it is only in that form that it became a visible and compelling public issue.

Over the past several decades, environmental concerns have increasingly tempered our enthusiasm for some of our own technology, but we still seem to share a collective faith in the "technological fix." That is, we firmly believe that whatever problems technology brings with it, yet more technology can be brought to bear to solve them, whether the issue is nuclear waste disposal, global warming, or the U.S. space program. In fact, biotechnology is sometimes marketed on this basis, as a fix for older problems generated by chemical-based technologies, as when genetic engineering is put to use in attempts to develop pest-resistant plants and thereby reduce our chemical pesticide dependence. Biotechnology is part of a bigger picture in which all technology continues to be seen as the source of solutions, as well as problems.

On the one hand, then, most agricultural biotechnology (or perhaps we should say biotechnologies) partakes of the aura afforded all science and much technology in American culture: these developments are to be welcomed, if not worshiped. On the other, however, some aspects of biotechnology—especially the ability to clone new individuals from an adult animal's cells—directly threaten our values and therefore become much more controversial. This is not to say that the influence of cultural priorities is always determinate, a fact that makes attempts at "spin doctoring" of related news a risky proposition. "Terminator gene" technology has been publicly defined as a threat to agrarian life and the economic viability of the individual family farm in ways earlier biotechnology that actually represented potentially similar threats (such as the development of hybrid corn seed) was not. Similarly, the development of genetically engineered bovine somatotropin, or BST, to boost dairy cow milk production provoked a lesser level of public reaction than the "terminator" technology but was still much more controversial than the earlier introduction of bioengineered rennet that took the place of that previously "harvested" from calves' stomachs. These kinds of differences can sometimes be explained after the fact—animal rights activists may have embraced biotech as saving calves, for example—but not always predicted.

Nevertheless, while it is not always going to be possible to predict when one set of values (such as our reverence for all things scientific) will prevail and when another (such as concerns about biological identity or economic autonomy) will not, to understand the development and impact of news about biotechnology would not be possible without taking into account the influence of the cultural context in which public opinion is formed—the more diffuse cultural matrix within which the climate of public opinion takes shape.

MASS MEDIA AND PUBLIC LIFE

This book seeks to explain public opinion about agricultural biotechnology in the United States today, with particular attention to the role of news media accounts. Media messages are not presented to readers and audiences with "blank slate" consciousness; seeing certain events as forming a connected series undoubtedly affects which cultural perceptions or frames of reference are brought to bear on their interpretation. We cannot understand the recent history of U.S. news and public opinion about agricultural biotechnology without understanding something of the entire sequence of events that created today's opinion climate. Much of this book therefore focuses on the news that has surrounded particular elements of that sequence during the 1990s. Other portions focus on the preexisting and emergent social, political, and cultural context in which that news has been interpreted. The cultural context as it exists in the United States becomes clearer by virtue of comparison, so comparative material from Europe, India, and elsewhere is also included. Finally, even though the public opinion data have consistently shown that people respond slightly differently to agricultural and to human medical biotechnology, as well as to biotechnology involving plants versus animals, these do not always represent clearly distinct categories in the minds of non-specialists, so we are concerned to some extent here with public responses to *all* biotechnology, not just *agricultural* biotechnology.

Fiction, as well as news, affects our cultural beliefs. As a general influence on the nature of the ambient cultural environment for public opinion forming around any given issue, fiction may in fact be even more influential than news. But this does not mean we need to be overly concerned with Frankenstein images and the mad scientist stereotype. Both of these are part of our culture to an extent and in ways this book will not be able to address. Biotechnology has at least the potential to dramatically impact human health, the environment, and the economy in both positive and negative ways. In the interests of democracy, we need to talk about this in public. Science fiction provides a unique and potentially valuable forum for discussing the promises and perils of modern biotechnologies where news itself may not because, by definition, news in this country actively avoids speculation—whether about the present or the future. We should not overlook the role of these popular culture products—nor assume their impact is universally negative.

However, this book is primarily about the news media, rather than other, perhaps equal or even more important, channels. It is intended as an issue-focused case study of the character and influence of U.S. news, understood by reference to the way it is produced and its relationship to culture and public opinion. Generally speaking, news accounts do not determine public opinion as much as is commonly assumed; they call our attention to certain issues, they invite (but do not actually dictate) certain definitions of situations, and they may in the process stimulate public debate. Most characteristics of news are not at all unique to topics in biotechnology, agriculture, or science (even though the cultural ambience is different for such topics than for news in other areas). In that sense, this book is somewhat parallel to other work on the emergence of issues that are less ambiguously political, such as Lang and Lang's (1983) study of agenda-building in the Watergate case and Gitlin's (1980) study of news coverage of Vietnam war protests. For this reason this book may be useful to those seeking a fuller understanding of media processes more generally, especially in relation to broader social and political trends.

The role of the news in such cases should not be understated, either. Even where the subject is science, U.S. news media do not just report events; they choose the events they report on and thus set agendas for the rest of us, and they suggest certain interpretations over others. Those choices and interpretations, while reflective of collective behavior within the general culture from which they spring, also privilege some positions—notably those of larger and more powerful institutions—over others, if only because of their heavy reliance on institutional sources for much of their ideas and information. This is not necessarily by design, but it is distinctly problematic for a democracy, especially one like ours in which a free and critical press is assumed to function to diffuse rather than to exacerbate uneven concentrations of power. Some within academic journalism are beginning to call for a return to so-called advocacy journalism that recognizes and supports the inevitably active role of the individual journalist in shaping public events, rather than denying these dynamics in the name of the objectivity ethic. As an academic study, this book is not an argument for a particular resolution, but it does illustrate some of the limitations of our existing system.

Most of the examples and analyses in this book are from the print news media, especially newspapers, although other sources of information are also considered, especially in the discussion of product labeling. Broadcast news, especially television news in all its modern forms from docudrama to talk show, is arguably more influential. But it is also vastly more ephemeral and therefore difficult to scrutinize systematically, as is even text-based Internet content. As a practical matter, post hoc considerations such as this must emphasize the printed word. This record by itself is generally coherent, if sometimes incomplete. As the Internet continues to expand its role as a source of news and information, the primacy of news printed on paper may continue to be eroded, a process with its own ramifications for both this study and the character of U.S. democracy. We will see this most clearly in considering how the so-called terminator gene controversy seemed to mark a U.S. public opinion watershed in late 1999.

2

Reinventing Milk

The Codex Alimentarius Commission (the joint food standards commission of the United Nations Food and Agriculture Organization and the World Health Organization) . . . has ruled unanimously in favor of the 1993 European moratorium on Monsanto's genetically engineered hormonal milk (rBGH).

—August 18, 1999, press release from Samuel Epstein, M.D., University of Illinois School of Public Health, under headline, "Monsanto's Genetically Modified Milk Ruled Unsafe by The United Nations" (Epstein 1999)

A Recent 'Press Release' (Aug. 18, 1999, S. S. Epstein) Alleging A United Nations Ruling That Milk From Cows Supplemented With BST Is Unsafe Is A Total Fabrication.

—Subhead of Monsanto Corporation press release (Monsanto Corporation 1999)

How can two such divergent interpretations of the same sequence of events come from apparently credible and well-informed sources? Bovine somatotropin (BST), also called bovine growth hormone (BGH), stimulates milk production in dairy cows. It is manufactured commercially by growing microorganisms genetically engineered to produce it, and it is marketed in that form by Monsanto Corporation under the trade name Posilac®. In this form BST is usually referred to as rBST or rBGH to identify its origins in recombinant DNA technology, or genetic engineering. The industry usually uses the term rBST rather than rBGH out of concern that the word "hormone" will be unnecessarily alarmist. This variation in terminology is one of the more pallid and tame reflections of one of the most acrimonious propaganda wars of the twentieth century, at least outside of wartime itself. It was a harbinger and a forerunner of an escalating confrontation between biotech's promoters and food consumers, and it set the stage for subsequent public reactions to other products of biotechnology. Industry messages contributed to this escalation by missing the mark of actual public concerns.

Milk from cows supplemented with rBST/rBGH is the first major U.S. food product to be affected by genetic engineering in such a way that the public noticed, so all subsequent introductions of biotechnology-based foods and medicines are going to take place in a climate of public opinion heavily colored by the debate surrounding rBST/rBGH. Monsanto has claimed that as much as 30 percent of U.S. dairy herds are rBST/rBGH supplemented. However, Monsanto representatives have also argued that rules permitting the labeling of rBST/rBGH milk are inappropriate because it is chemically impossible to tell the difference from milk not produced using rBST/rBGH. It therefore appears that it will also be impossible to verify their claim about how widely the supplement has been adopted. Regardless of its popularity with farmers, however, it is worth pausing to consider its impact on consumer evaluations and concerns.

Most journalists are ill-equipped to distinguish among the competing claims of expert sources who disagree, especially in such a technically and politically complex area, and in the case of rBST/rBGH it is clear that they make these distinctions at their peril. Presently, two broadcast journalists in Florida, who developed a series of investigative reports on rBST/rBGH use in local dairy farms for a Fox Network-affiliated station in that state, are suing their former employers for wrongful termination under Florida's whistle-blowing statute. They claim, in essence, that the station bowed to pressure from Monsanto in repeatedly asking them to revise (that is, soften) their story and finally canceled the series altogether, and that this kowtowing was in violation of Federal Communications Commission regulations. Since the journalists were (according to their version) fired for objecting to this unsanctioned action on the part of the station, their termination was in their view illegal. The existence of pressure from Monsanto through their attorneys is easily documented; however, the motives of the station executives involved are of course much more difficult to pin down. After numerous delays, the case finally went to trial in July 1999 but remains unresolved as of the time this book went to press (August 2000).

These journalists' story included information from one Dr. Samuel Epstein (quoted above) on a possible link between rBST/rBGH use and higher levels of a substance called Insulin-like Growth Factor (IGF) in milk. Epstein and others believe that IGF, in turn, can be linked to cancers of the breast and prostate. This link to cancer is arguably the most controversial of a range of claims that have been made regarding the safety of rBST/rBGH use. Other claims are less dramatic, but still disturbing. For example, by increasing milk production, dairy cow stress can be increased, possibly leading to increased incidence of mastitis and other infections, concomitantly increased use of antibiotics on the cows by dairy farmers, potentially increased levels of antibiotic residues in milk consumed by people, and thus potentially diminished effectiveness of some antibiotics in combating human disease. Yet other claims revolve around other specific health problems observed in treated cows. Not the least bit surprisingly, Monsanto appears to believe all such claims are also either overstated or erroneous.

In an August 23 communication presented as an Internet postscript to his August 18, 1999, press release, Epstein clarified his position that the recent decision by the Codex Alimentarius Commission not to set a maximum standard for BST "residues" in milk at that time in effect affirmed the rights of specific national governments (and the European Union) to decide independently whether or not to accept rBST/rBGH milk. It also avoided public debate on the wisdom of declaring that BST use is safe at *any* level. Had they agreed to a standard, in other words, the individual governments would presumably have been obligated to give up their right to set their own rules in deference to the internationally accepted ones. So for the Codex to *decline* to regulate this matter was, in Epstein's view, actually an action *in support of* countries such as Canada that have independently decided BST is a problem and set national policy on that basis. By prescribing no ruling to the contrary, the Codex allowed the 1993 European moratorium on BST milk to stand unchallenged. In the end, the Codex decision may have been as much a decision in favor of national sovereignty as one supporting either side of the BST controversy, and may well have been motivated as much by a desire to avoid protracted and unproductive debate as anything else, Epstein's attribution of this decision to "unarguable scientific grounds" notwithstanding.[1]

However, Monsanto probably should not claim this decision as a victory for its product either. Presumably, they would allege that a decision not to regulate must be a decision conceding the safety of rBST/rBGH milk. But appearances are deceiving here. It's not clear on exactly what basis they might want to celebrate the defeat of the proposal. If Codex adoption of standards in this area had occurred, BST's safety at *some* specified level, at least, would have been established, and participating nations would presumably have been compelled to accept milk that met the standards. Just to complicate matters, however, remember that Monsanto also claims rBST/rBGH leaves no detectable "residues" that can be distinguished from naturally occurring BST anyway. This has been the basis for their opposing the establishment of relevant milk labeling standards (see chapter 6). Had a standard been adopted, would the issue of assessing rBST/rBGH's impact on milk—and therefore the whole labeling controversy—have been revisited domestically as well? And how many dissenting voices on this controversy, up until then relatively muted within the United States, have been heard? All other Codex decisions made at this meeting were described as consensus-based; this one would not have been. A compromise based on maintaining the status quo in this area probably served Monsanto's interests well. It certainly avoided any public airing of differences over the associated trade issues.

Whether Epstein turns out to be right or wrong in this particular case, both scientifically and in terms of his perceptions of the reasoning beneath the Codex decision, the United Nations Food and Agricultural Organization (FAO) has also released a statement specifically disavowing his attribution of the decision to concerns over safety. In other words, whether for good reasons or bad, this sequence of events has effectively served to label Epstein as officially beyond the pale of established credibility, weakening the Florida journalists' own credibility and by association that of all critics of BST. Even if the extremist label is a reasonable characterization of Epstein (which of course remains unclear, and given the world trade stakes involved is an enormously politicized question), the fury over his claims both about a possible BST—cancer link and about the Codex proceedings have effectively diverted what little public attention has been directed toward these issues to those aspects most easily dismissed as representing extremist criticism. Mainstreaming of large-institution perspectives is greatly facilitated by the availability of adversaries who can conveniently be branded in this way. There is a public relations cost to Monsanto in this case, however; if Epstein is a nut case, why does the corporation feel such an obligation to discredit him?

This little story about a nondecision made at an otherwise largely invisible international meeting illustrates the complexity of the issues involved in this and other biotech controversies, in which much of what is at stake is public perception. But it represents just the latest minor skirmish in a long battle over scientific, medical, and economic evidence and its interpretation and management by Monsanto and other stakeholders. It is a minefield for journalists. The legal situation is so complex, the science so intimidating, and the calculated moves made by key players so consistently bewildering to outsiders, that it is, in fact, little wonder that the issues seem to be avoided in U.S. news media, where science journalists are rare and agricultural policy specialists are rare to non-existent. And what kind of interesting story could be made out of such technical material anyway, short of the ever-popular "cancer scare" motif proposed in Florida?

Ironically, though, it is probably the key stakeholders, especially Monsanto, who are most responsible for the current focus on health of what little popular news coverage there is of the rBST/rBGH controversies. Perhaps this strategic issue definition was selected because Monsanto and other stakeholders thought they were on the most solid ground with respect to the health arguments, or perhaps it was merely accidental, resulting from a misunderstanding of what critics were actually saying early on. Either way, the stage was set for this drama some time ago, when biotechnology's first major agricultural product was originally cleared for widespread use on U.S. farms. Food safety was not the original issue.

THE EARLY CAMPAIGN

On November 5, 1993, after years of political and regulatory negotiations, the Federal Food and Drug Administration quietly approved the use of rBST/rBGH in commercial milk production without requiring special milk labeling. After a moratorium that allowed time for a federal impact study to be completed, Monsanto Corporation began taking orders for the product on February 4, 1994. As the first product of genetic engineering to have a major impact on the U.S. food supply, rBST/rBGH had already been subject to much public scrutiny and speculation. Although its milestone introduction passed without much public comment in some areas, some reports indicated that schools in Berkeley and

Los Angeles planned not to buy artificially BST-bolstered milk, and a February 8 *New York Times* article reported that protests were staged in major cities around the country (FDA Warns 1994).

Part of the furor over rBST/rBGH was due to its status as the first significant food-related product of genetic engineering, with the potential to set a precedent for the regulation of all subsequent ones—and color future consumer reactions. But at the time human health was not the only, nor necessarily even the primary, concern. The executive summary to a federal impact study released in January 1994 noted that socioeconomic factors had never prevented the introduction of a new technology in this country, and in the end BST was no exception. At the time, however, significant concern existed that routine rBST/rBGH use would likely drive many small family dairy farmers (those with the most marginal profit situations) out of business. Since then, the argument has been repeatedly raised that less well-managed farms, not necessarily just smaller farms, would be the most adversely affected. Nevertheless, it is not accidental that small-farm dairy states in the upper Midwest and in the New England states were the places the BST controversy seemed the most heated, while larger-farm states such as Texas raised fewer objections. Its impact on the price support system for dairy products was also unclear, with some predicting the collapse of that system entirely. Others were more concerned that the product would simply benefit its corporate owners alone, with pricing set to absorb most of the potential profits from higher dairy productivity.

This chapter contends that some of the controversy over BST could have been avoided had its promoters handled their public relations activities more carefully. This would have meant more squarely approaching the issue of socioeconomic impact, generating a public dialogue that could have represented an opportunity for informed democratic debate rather than exacerbating the polarization of pro- and anti-BST positions. Such an outcome would have better served the interests both of rBST/rBGH's developers and of society as a whole. It would also have set a more meaningful and less contentious precedent for the future introduction of additional genetically engineered products in U.S. agriculture. While it may seem naive to propose that major agribusiness corporations should want to stimulate debate about controversial aspects of their own products, it may be useful to remember that all the fury over the ethics of cloning technology (chapter 7) did not result in its ban.

Rather, cloning now seems to have been quietly accepted. For whatever reason, rBST/rBGH continues to generate controversy, and consumer concerns about it around the world, while not acute, also do not seem to be abating.

The development of rBST/rBGH was underwritten at a cost of perhaps half a billion dollars by Monsanto Corporation and three other major manufacturers of agricultural pharmaceuticals (Eli Lilly, American Cyanamid, and Upjohn). Monsanto had the lead and filed the original application for FDA approval. Widely heralded and closely watched because of its precedent-setting potential, rBST/rBGH became an object of controversy while it was still being tested. Contemporary concerns centered on assertions that it might drive as many as half the dairy farmers in small-farm states out of business and that it would generate additional milk surpluses in an already glutted market, threatening the federal milk price support system, as well as concerns that its impact on health remained unclear. Wisconsin and Minnesota both temporarily banned engineered BST because of its projected economic impact; several major grocery store chains and many agricultural cooperatives were also reported to have refused to accept milk from BST test herds out of fear that consumers would perceive the milk to be impure or unsafe.

The public relations strategies that the developers of rBST/rBGH—most prominently the Monsanto Corporation—used when this controversy first erupted in 1989 are reflected in local newspaper coverage of that time. Although the product has since received full federal approval for commercial use in agriculture, it is likely that much of the controversy that both delayed this approval and set the stage for the introduction of other bioengineered agricultural products might have been avoided by the use of a different set of strategies. More to the point for purposes of this book, this example helps illustrate how large-institution perspectives tend to shape news accounts despite journalists' adherence to the objectivity ethic (the idea of balance). It also helps explain how the kinds of issue definition and agenda-building activities that were going on in the late 1980s and early 1990s indirectly influenced the later sequence of events. The picture of institutional action and response that develops illustrates some key principles of persuasion, risk communication, and public relation theory that help us weigh the extent to which Monsanto itself contributed to the public opposition that arose and how this opposition might have been met differently.

The storm of controversy and protest that the BST issue engendered seemed to take Monsanto and the other aspiring manufacturers by surprise. The product was developed in an apparently positive—or at least neutral—climate of public opinion. Louis Harris & Associates conducted a survey on behalf of the (now defunct) U.S. Office of Technology Assessment that was generally considered good news for biotechnology's developers when its results were released in 1987. *Chemical Week* (1987) reported the study under a headline that read, "Poll results may bode well for biotechnology." *Chemical & Engineering News* (1987) was more cautious, but began their report of the study with this lead: "Two thirds of the American public think that genetic engineering will enhance life for all." Another report emphasized that 74 percent of those surveyed were in favor of the use of genetic engineering to create better farm animals (Lacy, Busch, and Lacy 1991). Although the survey itself certainly uncovered some strong reservations about specific applications, the mood it seemed to capture was one of cautious optimism. The researchers found "no measurable decline in public optimism toward science during the 1980s" (USOTA 1987, 28), for example, and noted that the American public "approves nearly every specific environmental or therapeutic application" (USOTA 1987, 5) of genetic engineering.

The nemesis of Monsanto and the other developers was not, initially, general public opinion; it was resistance from dairy farmers and others concerned about socioeconomic impact. Although health and safety concerns, including those outlined above, were raised later on, this occurred in an environment in which rBST/rBGH's promoters and their market (which in the end is dairy farmers, not milk consumers) were already antagonists and the issue of the wisdom of widespread commercial use of BST was already on both media and public agendas. The potential impact of consumer fears (irrational or not) on dairy product consumption became one more argument in the arsenal of those in the dairy industry opposing BST's use on essentially economic grounds, and allowed them to join forces with consumer activists whose opposition was centered more clearly on health and safety issues.

The issue was not big in the national media at this time, however, and its possible health aspects were nearly invisible. The *New York Times* index for 1989 lists only four articles on BST, one of which was an editorial on how dairy farmers might best prepare for its economic impact, two of which concerned farmer opposition, and just one of which concerned rejections of BST milk by Ben & Jerry's Ice Cream and major grocery stores. Health issues were not the main theme of any of these articles. But the controversy had erupted much earlier in a few smaller dairy states, as can be seen by looking elsewhere than in the "prestige" press. The commercial Newsbank news indexing service, which includes nonwire service stories from smaller as well as medium-sized and larger papers throughout the United States, listed no articles specifically on BST for 1988 but fifty-five articles on related topics during 1989.[2] Of the fifty-five articles, nineteen were published in Vermont and twenty in Wisconsin. No other state accounted for more than three of the stories.

Key issues identified in the headlines of these stories fell into the following categories, which waxed and waned in prominence as indicated: dairy industry (farmer) reaction, which included all nine articles printed in April or May and one in June, plus two later in September; chemical industry position, which characterized one June article only; controversies over research, accounting for six articles spread about evenly from June through October; rejections of milk by dairy cooperatives, grocery chains, or other buyers, which accounted for ten articles, of which the first appeared in June, and peaked with six articles in August; political and legislative issues and activities, which began to appear in June but did not peak until four articles appeared in September, accounting for nine articles in all; disposition of test milk, the subject of five August-to-September articles; scientific background, the subject of two articles in September and October; public opinion, accounting for only a single September article; economic impact, the subject of two September articles; and other, including general overviews of the issue, which was made up of seven articles appearing in the September—November period.

In other words, dairy industry reaction was the number one issue overall, as well as the earliest identified, followed much later by public milk rejections and then by political/legislative activities, which numerically were the second and third largest categories, respectively. These three categories accounted for thirty-one out of the fifty-five articles, or 56 percent. Only one article was on public opinion, and none focused on health or safety per se, although this issue did come up regularly in the context of other discussions. Thus this coverage, like most news coverage, responded to the events and activities of key stakeholders (in this case members of the dairy industry in affected states) rather than concerns expressed by "the public." If anything, the coverage tended to de-emphasize rather than overemphasize consumer and activist reaction, especially in the initial period.

Jeremy Rifkin and his organization, the Foundation on Economic Trends, are widely recognized as leading opponents of BST use (as of other biotechnology). His tactics, as well as his positions, are often controversial. From the point of view of the chemical industry, all mention of Rifkin would probably be seen as sensationalistic. The existence of critics who can be labeled extremists and dismissed with relative ease is very handy for the purpose of demonstrating a nominal commitment to journalistic balance in a story essentially framed by an industry point of view, just as it is a handy device for shifting the media agenda toward those issues most readily discounted. Nevertheless, industry criticism of this early coverage as too heavily weighted toward anti-biotech perspectives is unwarranted. Monsanto or other industry spokespersons were quoted directly or indirectly twenty times in the fifty-five articles; spokespersons from the Foundation on Economic Trends only six times. Some reference to Monsanto or to another of the companies developing rBST/rBGH, or to the Animal Health Institute (an industry-funded "research" group), appeared in almost two-thirds of the articles (thirty-four, or 62 percent, of the fifty-five stories). The comparable figure for Rifkin and his organization was only 22 percent (twelve stories), and many of these were no more than passing references in concluding paragraphs, presumably tacked on to create balanced stories.

Using a short opposition-perspective reference or quotation in this way to wrap a mainstream-perspective story is very common and not at all confined to Rifkin or BST coverage. It became a very common way to tack ethicists' comments onto science-based cloning stories, as shown in chapter 7, and the practice of using this end-of-article position for provocative but less mainstream perspectives has become so conventional that it now seems to connote delegitimization all by itself. Be that as it may, Monsanto and other chemical-industry spokespersons were certainly more prominent in these news accounts than their best-recognized opponent and had many opportunities to comment on the issue from their point of view.

What did the industry use these opportunities to say? Strikingly evident were some rather extraordinary characterizations of opposition arguments. Monsanto representatives were variously quoted describing these as "ludicrous," "unfortunate," "ill-advised," "unfair," "misinformed," "unnecessary," "cynical," and "phony," and categorizing them as "hysteria" and "demagoguery." Monsanto was not always alone in attacking critics of rBST/rBGH technology; an Eli Lilly spokesperson was quoted as saying that the "whims of politicians" ought not to dictate research. The Dean of the College of Arts and Sciences at the University of Wisconsin–Madison, under fire for conducting industry-funded BST research, merely called fears "unsubstantiated." But Monsanto was the most conspicuous voice, and what they had to say was singularly acidic. This approach cannot have helped their position vis-a-vis the public. However, they also made active attempts to set BST in contexts that would have a more positive aura.

CULTURAL MOTIFS

Students of American culture's longstanding romance with science and technology will find many of the thematic links Monsanto and others conjured up for rBST/rBGH during this period to be familiar. These themes resonate deeply with American values and beliefs in this area. Five themes emerged in reviewing this coverage.

Equation with progress. Americans tend to believe that progress consists of the discovery of new science and technology. BST coverage compared it to other developments in agricultural technology that are generally recognized as benign: silos, computers, artificial insemination, and embryo transplants. rBST/rBGH was portrayed as the next step in the continuous march of progress that has brought us the agricultural productivity and economic prosperity we have today. Early objections to these other technologies were eventually dismissed; wouldn't the same thing happen with BST?

Inevitability. Time marches on, and the progress that comes with it cannot be stopped. Opposition is therefore futile and self-defeating. BST test milk would "definitely" be marketed through commercial channels; it might well be used by half of the dairy farms within ten years. (Here they were not too far off, if we accept their own one-third estimate as of 1999.) Closely linked to this contention and the one above is the argument that if farms were driven out of business by the introduction of rBST/rBGH it would only be the continuation of a long, inexorable process that had been in evidence for decades—one of the prices we must pay for our progress and prosperity, and one inevitable under any circumstances.

Value neutrality. The idea that science, even technology, is "value free" instead of reflecting the intentions and priorities of its creators is hotly debated among scholars, but from the corporate side of the BST debate, the whole issue appears settled. BST was described as "only a tool in the farmer's box." In this view, shared by quite a few in the United States, technology itself is a blessing that can do no wrong. If there are bad effects, they must be the result of misuse or misunderstanding, never anything integrally a part of the technology itself. A hammer can be used to build something or hurt someone; it is hardly the hammer's fault which way this goes. Arguments about social or economic impact can have no place in the BST debate, given this premise. They are defined as irrelevant; any bad effects must result from misapplication, not characteristics of the technology itself.

Development at great expense. The developers of rBST/rBGH argued that they would have been foolish to "spend a fortune" on an ineffective or unsafe product. By the time this argument appeared, it was beginning to look as though a fortune had in fact been spent on an unmarketable, if not necessarily ineffective or unsafe, commodity. Nevertheless, while this argument makes little real sense, it is a plausible extension of American materialism to value whatever costs a lot of money. Why would major corporations spend so much to develop something people wouldn't want?

Purity. Monsanto repeatedly emphasized that BST milk was a "safe and nutritious" product. The corporation has sometimes been criticized for choosing milk as the object of their first public relations experiment involving food-related genetic engineering. Milk, after all, is what mothers feed babies, and any suggestion milk is "tainted" is believed likely to disquiet consumers. Here, however, Monsanto trades on the association between milk and purity, perhaps hoping the glow will spread to their product rather than the other way around. It is doubtful they were successful.

The image of science on which Monsanto attempted to capitalize was that of a benign, progressive, and yet unstoppable force that could hardly be opposed on rational grounds. For the most part, as comparisons with responses in other parts of the world make clear, this image was not an inaccurate reflection of how Americans think about science, and we have been more willing than other cultures to apply this thinking to agricultural biotechnology in most cases. Consonant with the equation of technological advancement with social improvement that has been characteristic of American philosophy from the Revolutionary period forward, this unmitigatedly positive image is nonetheless no longer likely to be as good a reflection of the public mood as it had been in earlier decades. Three Mile Island, Love Canal, Chernobyl, Bhopal, the Challenger space shuttle—images of technology run amok are too ubiquitous in the contemporary world for this to be otherwise. Reacting to anti-BST forces as a fringe minority of present-day Luddites to be defeated through being discredited was probably not the wisest strategy, even from the point of view of narrow corporate self-interest, in a world in which the perception of technological risk is affected by recent history and takes into account social and economic issues (Hornig 1993).

An additional and quite prominent strategy that seemed, on the basis of this analysis, to be used consistently by large-institution spokespeople was shifting the blame to others. An Animal Health Institute spokesperson was quoted as saying that "dairy farmers . . . should pay more attention to . . . their own management practices," rather than blaming the chemical manufacturers for the product's anticipated economic impact.[3] Such a position could hardly be expected to win many allies among farmers. The media were also blamed. The *Burlington* (Vermont) *Free Press* reported that "Monsanto officials for months have argued that media attention has distorted the issue and scared off farmers." Elsewhere, the general manager of one dairy cooperative said of the media, "You are destroying our membership. You have aroused people's suspicion that this milk is something bad." And the head of the University of Wisconsin Dairy Science Department said that media use of the term bovine growth hormone (as opposed to bovine somatotropin) was actively intended to "sow fear and panic." While the media are often blamed for carrying bad news, these defensive stakeholder comments certainly did the stakeholders themselves more harm than good.

Monsanto representatives continued to stress an earlier FDA ruling that BST milk was safe for humans. "If you can't trust the FDA," Monsanto asked, "who can you trust?" This tactic surely underestimated the current level of public mistrust of regulatory agencies, but perhaps more importantly, if this approach had any impact at all it would probably have served more effectively to shift the agenda to health concerns than to diffuse public concern altogether. Meanwhile, the FDA, a large-institution stakeholder in its own right, defended its position and claimed that associating BST use with negative human health effects was "a gross distortion of scientific fact." Possibly the industry felt itself to be on more solid ground with respect to health concerns and used this strategy in an attempt to divert attention from economic issues. Nevertheless, and even if Monsanto and the FDA were completely correct, the health-effects agenda continues to haunt discussion of rBST/rBGH use to this day, now elevated to the international stage.

Relegated to the background by this sparring over the credibility of health-related data is the issue of whether objections to a technology based on its social and economic implications (as opposed to health, safety, or effectiveness concerns) could ever constitute valid grounds for regulating a technology. On this issue the perceived self-interests of pro- and anti-BST groups seemed to converge. Just as BST stakeholders may have felt more comfortable on the well-defined territory of policy based on health alone, so BST's opponents may have felt that stirring up public fears about health effects might be easier than mobilizing public opinion on other grounds, and they may have been right. But while occasional journalistic accounts during this early period featured worried consumers, both individually and collectively the general public was far less prominent in the news than claims about consumer fears made by the big-institution players. In other words, public reaction was not itself driving media accounts.

INSTITUTIONAL INTEREST AND PUBLIC RESPONSE

Lay people's evaluations of risk may be higher for issues presented (or framed) in terms of their sociopolitical dimensions than those more narrowly formulated in terms of direct risks to health (Hornig 1990). This may simply be a matter of the higher degree of uncertainty attached to social impact considerations. Regardless of the reason, on this basis the self-interest of the biotechnology industry might at first appear to lie on the side of a strategy of concentration on the health issues debate. However, this is true only superficially and only in the very short term. Just as the media-cultivated illusion of infallibility that surrounded the National Aeronautics and Space Administration became its nemesis when the space shuttle Challenger exploded, social impact issues for biotechnology are not likely to go away, and sidestepping them now is only likely to exacerbate future public relations problems. For rBST/rBGH, the economic issues were raised early on by the dairy industry and therefore were especially unlikely to pass unnoticed more generally.

While biotechnology is not the only area in which journalists have been blamed for creating controversy by covering it (and, in fact, agenda-setting theory does support the assertion that this is possible), science journalists have more often been subject to the opposite critique, accused of serving as publicists for the scientific community. Identifying strongly with the institutions and interests they are assigned to cover, these journalists may see their job as selling science to the lay community without enough attention to the costs, risks, uncertainties, and controversies associated with science and technology (Nelkin 1995). Journalistic ethics (the criterion of balance) dictates that opposing sides cannot be ignored, even though the delegitimization of one side or the other can be accomplished in subtle ways in an otherwise "balanced" story. But journalists (including science journalists) rarely *create* news out of nothing.

Most news stories are source generated; in the case of science and environmental news, this means that stories arise because individual scientists or—as is much more likely—university or governmental public information officials, representatives of activist groups, or public relations specialists from the corporate world initiate them (for discussion see Gandy 1982). Scientific agencies and science-based corporations represent vested interests with increasing stakes in the formation of public policy under the influence of accelerating competitive pressures (Dickson 1984). And institutions within the biotechnology community seek to frame public perceptions of policy issues in ways they feel will be to their advantage as well (Plein 1991). Reporters are supposed to remain independent of their sources, and most journalists probably try to do their jobs in an ethical manner. However, they are in fact heavily dependent on sources for information as well as story ideas, especially in technical areas. Coverage of the BST controversy is certainly no exception and serves to illustrate the way large-institution dominance of the news plays out.

Most news coverage reviewed for this chapter was balanced in the sense of including opposition perspectives. The evidence does not support contentions that the news coverage either ignored industry's claims or sensationalized those of consumer activists. Rather, these news reports tended to rely heavily on university scientists engaged in relevant research, whose information is generally considered objective by definition, as sources. Like journalists, scientists are also largely ethical, and their stakeholder interest on these issues is not the same as that of agricultural biotechnology's corporate developers. However, scientific reputations and careers, as well as scientists' incomes, can become bound up in public controversies about their work. Furthermore, scientists who do research about biotechnology do not have the same opinions about related policy issues as equally knowledgeable scientists who do not do research about biotechnology (Priest and Gillespie 1999). If the media coverage had a bias, then, it was more than likely a pro-biotechnology one on the basis of sourcing patterns. Available data tend to confirm this.

If anti-technology media bias is not to blame for a poor public reception for rBST/rBGH, then what is? Jerry Caulder, president of the Industrial Biotechnology Association, reportedly blamed the reaction on public scientific ignorance in a discussion with *Science* magazine (Roush 1991), and similar arguments have consistently been raised from various representatives of the biotech industry and the relevant research community. However, little evidence supports the claim that increased knowledge results in increased pro-biotech sentiment, and some evidence supports the opposite conclusion (Gaskell et al. 1999). While providing information and education may change attitudes, especially on a longer-term basis, ignorance is not an adequate explanation for negative public reactions. The agenda-shifting and delegitimation strategies of corporate, governmental, and scientific stakeholders—who, as large-institution voices, disproportionately control the character of news accounts—may be as much the cause as anything else.

The public form of this controversy was at odds with the two-way communication model characteristic of effective public relations (Grunig 1989), as well as with the ideals of democratic debate in which individual citizens are empowered to participate. The biotechnology industry seems to have either misread its audience or chose a strategy designed to shift the discussion to another set of issues (health rather than socioeconomics) that it perceived as "safer." Whatever their motivation, this choice demonstrated neither awareness of nor sympathy with the actual concerns of the dairy community, the source of the earliest criticisms. Rather than addressing socioeconomic impact issues directly, most large-institution messages either ignored these questions or sought to discount them as appropriate subjects for public debate. Instead, the industry relied on ads such as one headlined, "You've had BST and cookies all your life," attributed to cosponsors Eli Lilly, Monsanto, and Upjohn (Roush 1991, 34).

By so doing, these institutions lost the persuasive advantages that authentic two-way communication (see O'Keefe 1990) might have provided. In the end, even the FDA recognized the fine line between education and propaganda by asking Monsanto to stop promotion of rBST/rBGH pending final approval for commercial use. The extent to which these strategies might have undermined the chances of some of these institutions for maintaining long-term public support is something we will never really know. However, in the shorter term at least, this approach was still far more effective in the United States, a nation characterized by a legally free but economically constrained press system that is heavily dominated by a single electronic news service (the Associated Press), than it was elsewhere in the world. This may well reflect the relative power of mainstreaming effects for U.S. news. Recent public relations efforts on the part of the biotechnology industry have been more sophisticated, addressing a broader range of public concerns more directly.[4] But the challenge of selling biotech products on both United States and world markets may require more than producing better persuasive messages. Without authentic opportunities for democratic debate and public choice, it seems unlikely that resistance will diminish.

NOTES

1. The relevant FAO press release, numbered 99/41 and dated July 4, states only that a decision on BST limits would be postponed "until a consensus is reached" and that an intergovernmental task force to expedite the development of standards for biotechnology-based foods had also been established. Reasons are not stated for either decision.
2. The author or an assistant read and abstracted for further analysis all fifty-five articles listed under these keywords: milk, dairy, bovine growth hormone, and bovine somatotropin.
3. Quotations in this paragraph and the one following are also selections from the data described above.
4. See, for example, the information from the Council for Biotechnology Information at <http://www. whybiotech.com>.

3

World Reaction

A common myth used by Monsanto and the Biotechnology industry is that without genetic engineering, the world cannot be fed. However, while biotechnology is projected as increasing food production four times, small ecological farms have productivity hundreds of time higher than large industrial farms . . . .

—Dr. Vandana Shiva, director, Research Foundation for Science, Technology and Ecology, New Delhi; statement prepared for International Conference on Women in Agriculture, June 28-July 2, 1998, Washington, D. C. (Shiva 1998)

The American Corn Growers Association recently predicted that growing uncertainty over market acceptance, both foreign and domestic, of genetically modified organisms or GMOs would result in a dramatic decrease in the use of genetically modified seed this coming season (http://www.acga.org). Requirements from some countries that genetically modified crops be segregated from others and the degree of economic concentration in the agribusiness industry that could result from a few companies controlling the supply of patented seed corn have been among their stated concerns. The association, which claims to represent the interests of corn producers in twenty-eight states, has recommended to its members that they look at the option of planting non-GMO crops which are likely to demand premium prices in the near future. While it is difficult to gauge whether this is becoming the typical reaction of U.S. farmers, it closely parallels dairy industry reaction to the introduction of rBST/rBGH. Again, however, the reaction to genetically modified crops and food products seem to have taken the biotechnology industry by surprise—despite the fact that their experience with BST might have predicted such an outcome. Once again, the debate centers less on health and safety issues than it does on economic and political ones. It now rages around the world with considerably more strength than it had until recently garnered in the United States.

Reaction in what is sometimes called the "developing" (or "less developed" or "Third") world has been the most clearly focused on the socioeconomic and geopolitical dimensions. Development theorists continue to promote science and technology as the high roads to prosperity, health, and well being throughout the world, and of course there is considerable truth to this assertion. The "green revolution" in agriculture of the 1970s and beyond increased food yields dramatically through the use of improved crops and farming techniques, enhanced by the application of technical know-how from the developed world. While lingering questions about the sustainability of capital- and technology-intensive farming among the more economically marginal nations of the world persist, few doubt the contribution of Western science and technology to staving off widespread developing-world starvation.

Furthermore, agricultural biotechnology has been heralded everywhere as the source of the next such "revolution." No less than in the developed world, however, the long-term environmental and economic effects of this forthcoming revolution are difficult to envision, let alone foretell. Corporate interests do not thrive by engaging in non-remunerative humanitarian activities, and the worldwide marketing of genetically engineered crops and livestock is inextricably tied to corporate interests. Engineering the acceptance of biotechnology-based agriculture around the world is not simply a matter of upgrading education or knowledge for altruistic purposes. It is also a matter of opening up new markets for biotech's industrial developers. In other words, it is the colonialism of the new millennium.

The old colonial division between capital-rich and natural resource-rich areas of the world remains, and the less-developed but resource-rich world is under increasing pressure from the capital-rich, developed, industrial world to manage those resources in a sustainable way. This pressure comes from a developed world threatened by the notion that its population might have to radically adjust wasteful lifestyles to environmental and economic realities. Simultaneously, developing-world nations are also under pressure to open themselves up as markets for Western products, including both biotechnology and scientific expertise. Of course, if people in highly populated but less developed countries such as China adopt materialistic Western lifestyles, as they appear eager to do, sustainability will be impossible to achieve. In a perpetuation of this old division between "have" and "have not" countries, Western companies have been widely accused of wanting to patent and market, in unfair and exploitative ways, the knowledge and traditional genetic resources (plant varieties, livestock strains) of people living in less developed countries around the world. However, responses to today's biotechnology are based on cultural beliefs, values and priorities, as well as concern about economic effects, and these cultural factors have produced widely differing reactions that are not necessarily reducible to differences over economic issues. And they are not at all limited to the less-developed world.

PRESS AND PUBLIC OPINION IN THE WEST

The comparison between Europe and the United States alluded to earlier helps establish that these differences in reaction are not simply a matter of differences in levels of scientific ignorance versus education. Objections to GMOs and controversy over IPRs (intellectual property rights, that is, patent rights to genetic and other biological and biomedical information) raised in the developed countries of Europe and in Canada are hardly the result of populations who lack sophistication through inadequate exposure to science. Rather, the willingness to accept these new technologies as benign probably depends a great deal more on factors such as faith in the adequacy and integrity of regulatory agencies, the wisdom of the scientific community, and the ethics of businesspeople than it does on scientific knowledge or understanding of specific potential risks and hazards per se (Sparks, Shepherd, and Frewer 1994; Priest 1995). These responses also depend on levels of concern about the risks of both environmental degradation and economic exploitation, which in turn are influenced both by cultural values and social experience. And economic and environmental concerns entail issues of social and environmental justice, not just net gain or loss. All of these issues are treated somewhat differently in political and media environments that also differ.

Public concern about "mad cow" disease in Great Britain has sometimes been cited as an example of how irrational the British public is about food safety issues, with the tabloids usually taking the blame, just as public fears about genetic engineering there have been blamed on

"Frankenfood" headlines. The conservative sectors of the British press have traditionally restrained from some forms of criticism and the British press as a whole is subject to significantly more restrictive libel laws than exist in the United States, which has certainly dampened their enthusiasm for controversy up until recently. The influence of the British tabloids is relatively new, which may partially account for their being blamed for what is seen as public hysteria over various contemporary food safety issues. But the U.K. government eventually conceded the dangers to humans associated with "mad cow" disease, or bovine spongiform encephalopathy. Concerns about the transmissibility of the disease to humans (where it takes a somewhat different form) generated a prompt European Commission ban on British beef imports; more recently, in the U.S, restrictions even on blood donations by those who have spent measurable amounts of time in Great Britain have been initiated. And it was discussion of "mad cow" issues in a 1996 segment of the Oprah Winfrey show that caused a group of Texas cattle growers to sue the talk show hostess under the state's False Disparagement of Perishable Food Products Act for undermining their market. The suit was unsuccessful.

Did members of the British public simply confuse bovine spongiform encephalopathy (BSE) with bovine somatotropin (BST) or bovine growth hormone? Either way, it seems likely that the earliest public concerns about BSE in Britain were as much a result of repeated official reassurances that nothing could possibly be wrong, as they were a function of sensationalized news coverage. Yet officials chose not only to discount such concerns as "merely a panic," but to blame them on the tabloid press (Brookes 2000). Whether or not the similarity in abbreviations helped to link the two controversies, erosion of public faith in regulatory integrity was a very probable connection. Public trust is easily eroded and regained only with great difficulty, if at all.

Patents on genetically modified organisms, genetic information, and genetics-related processes also became public issues in Europe in ways they simply did not in the United States, where the extension by Supreme Court decision of existing patent protections to genetically novel organisms was implemented with little apparent furor.[1] The European Parliament and European Commission actively debated the wisdom of patenting life forms and genetic information for an extended period of time on multiple grounds, including ethical considerations (Sterckx 1997). This was a debate confined to the activities of a handful of activist groups in the United States, where owning this kind of intellectual property has engendered little or no serious or sustained public discussion (with the exception of fairly regular mention of turf skirmishes between public and private interests engaged in collecting the rights to various segments of the human genome).

A series of public opinion surveys known as the "Eurobarometer" has made possible comparisons across Europe and across time of public perspectives on biotechnology (Gaskell, Bauer, and Durant 1998). This research has documented a logic focusing on moral issues, low levels of trust in regulatory agencies, and at most a modest relationship between knowledge and support. Across the seventeen countries participating in the 1996 survey, a majority of the respondents (52.2 percent) tended to disagree that current regulations of biotechnology provide enough protection; over twice as many believe that this technology is likely to create dangerous new diseases (69.8 percent) than that it is likely to substantially reduce world hunger (34.2 percent). This was true despite a press that was more likely to print stories on biotechnology's benefits than its risks and in which, more generally, positive stories were more common than negative ones. Universities and industrial interests dominated the European news coverage, just as they did the coverage in the United States. These facts illustrate the limits of conceptualizing public opinion as a simple function of the impact of media messages. Nevertheless, neither can the U.S. press be entirely acquitted of contributing to "mainstreaming" effects and de facto suppression of both public debate and public dissent. Effects of this type are likely to be subtle.

Does the minimization of dissenting views and the domination by large-institution perspectives in press accounts actually lead to a spiral of silence effect whereby, in public, only prevailing or consensus views (or those so perceived) are expressed? If so, just how likely is the diversity of opinion that might lurk beneath the surface to erupt later on? These are difficult questions to answer. European press coverage of biotechnology has been more heavily dominated by university and industry actors in some countries, such as the United Kingdom and Germany, than in others, such as Italy and France (Durant, Bauer, and Gaskell, eds. 1998). But links between the nature of press accounts and public perspectives on biotechnology and food safety are difficult to establish; all vary widely from country to country. One complication is that different countries within Europe may be at different stages of the debate, with the more heavily industrial countries likely to feel the influence of biotechnology stakeholders sooner. Another is that the importance of farming and general agribusiness to the economy of each country, and the respective levels of reliance on advanced technology within agriculture, also vary, as do the levels of consumer, farmer, and environmental activism.

Despite these challenges, it certainly seems reasonable to propose that public opinion that emerges as a result of public debate is likely to be more robust, as well as a more democratic foundation for public policy. In several areas of the world, including Canada and Australia, efforts to encourage citizen debate and deliberation through the use of "consensus conference" projects are currently underway. In these efforts, which take various forms, ordinary citizens are invited to participate in a structured debate on issues involving biotechnology in the hope of setting public policy in this complex area in a democratic way without ignoring or dismissing either public opinion or scientific evidence. The idea is that the citizen participants will engage in fact-gathering and reasoned deliberation, although sometimes relying on expert testimony rather than simply offering off the cuff reactions to something novel, unknown, and perhaps feared. In other countries and for other individuals, the media provide a substitute for this kind of engaged debate.

WORLD PRESS ISSUES

The processes of media agenda-building and public opinion formation in the rest of the world are in broad outline often very similar to those in the United States, but they respond to a different set of cultural, economic, and sociopolitical conditions (Hachten and Hachten 1999). Certainly differences in press systems contribute to global variation in public opinion; although the media are not the only institution that sets the public agenda, they are important participants in the agenda-building process. It is ironic, however, that the press in much of the rest of the world, which almost without exception enjoys a much lower level of legal protection than does the press in the United States, is often more diverse, at least outside of those countries with the most highly authoritarian or totalitarian governments. In the United Kingdom and much of the rest of Europe, for example, most newspapers have an identifiable political perspective, and news reports are not expected to be objective in the sense of being devoid of political meaning—despite the tendency for larger-institution perspectives to dominate just as they do in the United States. Even more ironically, it is just this kind of a partisan press that the U.S. First Amendment guarantees were originally written to protect, rather than the kind of commercialized but politically neutral press that we have today in which most news markets are served only by a single paper. It may never be possible to prove that expressed diversity of opinion changes the extent and character of public debate, but our legal-political system is based on this assumption. It was to preserve diversity and the right to voice minority opinion that press freedom was protected.

However, the partisan press began to be lost in the United States with the advent of advertising-supported, mass-circulation papers in the nineteenth century. This trend was exacerbated by the development of the wire services and various forms of electronic communication, including television, clear up to and beyond the advent of the Internet and the World Wide Web, arguably the first major electronic communication technology that serves to increase rather than reduce diversity of communication vis-a-vis social and political controversies. The advent of cable television was also expected to have an effect of this kind, but it failed to live up to its anticipated potential to increase diversity of views. Producing television programming is expensive, and experiments with community-based TV broadcasting have been disappointing, in part because the result was programming that could not compete for audiences with slickly produced national fare. Black and Hispanic channels now exist in many markets alongside Anglo ones, albeit at a likely cost in terms of both minority audience share and minority-oriented programming on the nonminority stations. Religious programming is reasonably common, but political and social diversity in messages remain rare, although it is seemingly more common in radio, where production is cheaper and costs can therefore be met with smaller audiences.

This kind of mainstreaming has not yet taken over the Internet, but there are signs that it may before long. Because of the time investment and skills required to produce a Web site that can compete in terms of glitz and gimmicks with those of major institutions, the latter are beginning to seem more dominant. Nevertheless, as later chapters will show, there can be little doubt that Web electronics is facilitating the broadcast of alternative points of view in ways TV did not, and contemporary debates about biotechnology are an excellent example. Although the extent to which the structure of the U.S. press system contributes to mainstreaming and the subsequent spiral of silence effects on public opinion—let alone its relationship to a political system in which there are only two parties and those are increasingly hard to distinguish—is difficult to assess, it is certainly real. U.S. surprise that the rest of the world is not always so eager to glean the benefits of our new biotechnologies is itself a rough indicator of the monolithic character of U.S. opinion (as expressed in public) in comparison to the situation elsewhere on the globe. Were the publicly expressed viewpoints in the United States not so homogeneous in assuming that these technologies would be quickly accepted everywhere, the present level of global controversy over biotech might not be coming as such an apparent surprise to its promoters.

TRADE NEGOTIATIONS AND "THIRD WORLD" ISSUES

At the "Rio Earth Summit" of 1992 (the United Nations Conference on the Environment and Development held in Rio de Janeiro), biodiversity emerged as a major issue, and the proposed Convention on Biodiversity (which then-U.S. President George Bush refused to sign at the time) addressed biotechnology and intellectual property rights in this context. On the one hand, biotechnology can be seen as an approach to conserving biodiversity because it promises to enable science to create agriculture that is in many ways more environmentally friendly. On the other hand, biotechnology as actually practiced may have the opposite effect. Crops engineered to be herbicide resistant make it possible for more herbicides to be used, herbicides that could have impacts beyond the fields on which they are applied. Crops engineered to be pest resistant may have unintended and unpredictable consequences on complex ecological systems as well. Either kind of crop could cross with wild varieties. And the dependence on expensive seed may encourage the kind of larger-scale, more technologically intensive agriculture widely practiced in the developed world, where the planting of vast tracts of land with identical crops eliminates natural diversity and therefore may carry an increased risk of agricultural or environmental catastrophe. These high-technology practices, in turn, are also linked to the survival (or extinction) of traditional human cultures organized around other types of agriculture, as well as to the fates of endangered species and over-stressed natural resources around the world. (For discussion of how these dynamics play out in Central America, see Utting 1993;

Eden and Parry 1996 discuss the social and environmental impact of land degradation in tropical environments more generally.)

Over the last decade, in part because of the new genetics that allows scientists to identify and sometimes transfer properties from one species to another, interest in species known to traditional peoples around the world has continued to explode. Meanwhile, genetically engineered cultivars (in addition to genetically engineered foods) are suddenly being marketed globally, raising new and difficult policy questions in the process. What, for example, would be the impact on the developing world, where farmers have been saving seed from one season to the next for thousands of years, of internationally recognized plant patents that might restrict this practice? Who should benefit from the commercialization of naturally occurring properties or substances—those who conserved these species, or those who discover new ways to make use of them? Voices within the scientific community (including a 1993 article in the widely read science news magazine *Nature)* urged the scientific community to take up the challenge of resolving these questions because of their significance for how genetic research would be conducted. However, they remain incompletely resolved to this day.

Just as the struggle over bovine somatotropin set the stage for subsequent polarization on the use of genetic engineering in producing U.S. agricultural commodities, the "Uruguay Round" of talks that took place between 1986 and 1994 on the provisions of the General Agreement on Tariffs and Trade (GATT) in some ways set the stage for subsequent debate about international intellectual property rights related to agriculture, plants, and animals and the right to refuse GMO imports despite free trade provisions. Approved by the U.S. Congress in December 1994 following twelve years of talks among more than 100 nations, the GATT bound signatory nations to cooperative protection of patents and copyrights around the world. Exactly what this will mean for commercial biotechnology and the interests of the developing world remains murky. The GATT provides for extended patent protections to microorganisms, nonbiological and microbiological processes, and animal and plant varieties. Its provisions continue to be controversial because of their association with issues of economic equity, environmental integrity, and cultural survival. The subsequent collapse in late 1999 of the World Trade Organization talks in Seattle, Washington, can reasonably be taken as an indicator of the extent to which such issues remain unresolved.

INDIA: A CASE STUDY

The interactions among culture, political interest, and the institutional characteristics of the press in different national contexts are complex, and teasing out individual causes is not always practical or even possible. Nevertheless, one of the best ways to understand how the U.S. press responds to issues is to look at these responses in a comparative context. The extensive published comparisons with the European press that have been completed have already been briefly described. But the contrasts among systems with fundamentally similar sociopolitical traditions, as revealed in those studies, are not always so dramatic as those between the biotechnology discussions in the U.S. press and those in the press in the less developed world. A quick comparison of GATT and related issues as they appear in the pages of the *New York Times*, the U.S.'s elite "newspaper of record," in comparison to accounts in the *Times of India* helps illustrate this.[2]

India provides an especially interesting comparison because controversy over agricultural biotechnology is significant there and because, while India is rapidly developing and has a quickly expanding scientific sector of its own, it retains its reliance on traditional small-scale agriculture that is not technologically intensive. Also, the constellation of religious and philosophical traditions in India is distinct from that in Europe and North America. Hinduism, India's dominant religion, holds some plants and animals to be sacred and regards the whole of the universe as one; along with the Hindu beliefs in karma (personal responsibility) and reincarnation, these values are consistent with a strong environmental ethic, respect for biodiversity, and a cautious approach to tinkering with biological processes and relationships. It is therefore not surprising that India has been one of the places where substantial and persistent resistance to agricultural biotechnology has been seen.

The news comparison data presented below focus on the period from April 20 through July 20, 1994, the three-month period immediately following the final GATT negotiations in Morocco.[3] Material used in the comparison consisted of all stories about issues involving plants and animals; these included articles on GATT, natural resource management, and IPRs. Altogether, the comparison involved fifty-five stories from the two daily papers. The *New York Times*, as is well known, boasts a readership that includes the intelligentsia across the nation, not just New Yorkers. The paper ran a total of just eleven stories about these issues during this period. About two-thirds of these stories could only be classified as general in nature; none of the articles addressed political or economic issues primarily. A single science story appeared during this time, and the remainder discussed various impacts of other kinds. Almost all of the stories were neutral in tone; almost three-fourths were on natural resources issues, with just one article on GATT itself and two on IPR issues. The stories tended to be analytical rather than news items, ran on average just under seventeen column-inches in length, and used an average of five sources each, suggesting they were well researched. But their scope was unquestionably restricted.

The *Times of India*—like the *New York Times,* a leading English-language daily—carried forty-four different stories during this same period, or exactly four times as many as its U.S. cousin on these same issues. This coverage consisted of somewhat shorter articles (averaging 13.5 column-inches) using fewer sources (an average of about one and one-half per article), and about two-thirds were straightforward news pieces rather than analysis. Some of these differences probably reflect differing news traditions and editorial tastes, rather than differences in treatment of this particular set of issues. Nevertheless, the total amount of coverage in column-inches is still striking, with more than three times as much in the *Times of India* as in the *New York Times*. The majority (57 percent) of the *Times of India* stories were about the GATT, with IPR and natural resource issues in roughly equal proportions accounting for the rest. Even so, this meant that the *Times of India* carried as many IPR stories as the *New York Times* carried stories on the entire constellation of issues in total.

In other words, GATT was much bigger news in India than elsewhere in the English-speaking world. And only about three-fourths of the *Times of India* stories could be classified as neutral, with nearly one-fifth negative and a handful positive in tone. Many of the *Times of India* stories were about the political dimensions of these issues, a topic that did not appear in the *New York Times* at all. In India, GATT-related issues became visible and political dramatically more than elsewhere, to judge by the characteristics of these news accounts, and this debate took place in public. While it would not be reasonable to identify the nature of the news in the two different countries as a primary *cause* of the differences in response, this contrast is clearly linked to differences in the agenda-building processes in the two nations, in which news organizations in each case are active—not passive—participants.

Looking at this coverage over time is also revealing. The *New York Times'* two GATT articles were published toward the middle of April and its lone IPR article appeared in early May. Natural resource management issues were covered consistently from early April through the middle of June, although most of these articles appeared before mid-May. Despite early attention to GATT issues, then, interest quickly died and was never as strong as that focused on natural resource questions. It is likely that environmental interests in the U.S. contributed to the emphasis on natural resource management, but GATT itself was not considered much of a news item. The existence of early coverage suggests this was not a matter of journalists' being unaware of the events; rather, the issues simply did not become politicized and were not pursued over time.

The *Times of India* actually had no GATT coverage before mid-April, from which point on it consistently dominated coverage of natural resources and IPR, with more than one article appearing eight out of the ten weeks from mid-April into the latter part of June, when the coverage finally stopped. Natural resource issues were the subject of only a single mid-May article until this topic started to appear more consistently in early June. IPR discussions were scattered throughout the period from late April on.

In short, where the *New York Times* had focused on natural resources and largely ignored the political and economic aspects of international policy at this time, the *Times of India* was part of an agenda-building process in which these aspects became politicized, attention was focused on GATT, and IPR issues were part of the public debate. The *New York Times'* well-known slogan is "All The News That's Fit To Print"; apparently much of the GATT controversy just wasn't. The result was relative invisibility in the U.S. press for emerging concerns in the developing world, feeding our complacency about world opinion.

Also in 1992, the U.S.-based W. R. Grace Company obtained a U.S. patent on its process to stabilize a traditional, naturally occurring, substance with pesticidal properties that is found in Indian neem tree seeds. This event erupted into a controversy that seemed to serve as a catalyst for debate in India and attracted the attention of the outside world to the potential impacts of world trade rules on agriculture. The neem controversy undoubtedly fueled the politicization within India of discussions of the GATT. Neem is widely known and used in India and elsewhere, not only as a pesticide but also in a variety of medicinal preparations. The patent by Grace engendered protest because it was widely believed that if recognized within India this patent would prevent traditional Indian farmers from using the readily available seed in the ways that they always had. While the patent was later reported to have involved the stabilization process and not the seed itself, concerns lingered regarding the morality of a multinational corporation profiting from neem with no specific benefit to the people from whom it was adopted, concerns that persist to this day. The impact of radically expanded external markets for neem on its local availability and cost remains difficult to determine.

GATT rules specifically require India (and other participating nations) to modify domestic laws to make them compatible with patenting on an international basis, which certainly does imply new domestic restrictions over neem use in response to commercial interests within the U.S. even if traditional usages were entirely unchallenged by the new rules. Thus the neem controversy became inseparable from the debate over the GATT rules, eventually resulting in the formation of a coalition of U.S.- and India-based activist groups for the purposes of opposing both the U.S. patent and the proposed revision to India's domestic patent laws. Once again, however, U.S. news media showed little curiosity about these issues, despite their apparent relevance to U.S. agribusiness interests. Persistent resistance to IPR "reform" continues to surprise us.

INTERNATIONAL NEWS AND CORPORATE INTEREST

While the media's direct influence on public opinion is easily overstated, news accounts are more likely to be influential under circumstances in which news audiences do not have other sources of information and are therefore most dependent on media accounts of events (DeFleur and Ball-Rokeach 1989). This certainly describes the situation with the GATT accords, both because of their technical complexity and also because of the international scope of the events, which far removes them from the day-to-day experience of almost all of us. Under circumstances of this type, the media are likely to play their most actively influential role in selectively framing or interpreting events and calling them to our attention (their "agenda-setting" function). These processes, including the treatment of alternative voices and interpretations, are sensitive to the cultural and political environments in which news is produced.

For this reason, looking at how a global issue is treated in the international press gives us a unique window on how cultural perspectives vary. In India, given ready public identification with the neem controversy, the public debate was sustained; in the United States, under other circumstances, it simply was not. However, the United States press was not necessarily just responding to local interest in these issues. The ways in which Western news treats various perspectives on biotechnology—the processes of legitimization and mainstreaming with which this book is primarily concerned—also reflect the old power differential that arose globally during the colonial era between the developed and the developing worlds. While it is perhaps reasonable and not particularly sinister that news produced in the developed world should reflect a developed-world perspective, critics have regularly pointed to the domination of world news by wire services and other large media organizations reflecting a Western orientation as a new form of imperial conquest.[4]

The newest colonialism is economic, not directly political, and the terms of world trade will determine who are the winners and losers in this "New World Order." The United States was never an old-style colonial power on the scale that many European nations were, but it is very much an imperial power today militarily and especially in terms of global trade and resource extraction patterns. And biotechnology (as science, as technology, as agriculture, and as medicine) is connected to a great deal of both what it is we have to trade (from food to technical expertise) and what it is we are most interested in extracting (often, genetic resources, whether pesticide-resistant plants, livestock strains adapted to difficult environments, or knowledge of medicinal properties with commercial potential). U.S. media are very unlikely to be engaged in any kind of active conspiracy to support U.S. trade interests; nevertheless, U.S. media corporations share many of the interests of U.S.-based multinational agribusiness corporations and other trade-oriented groups. So it should hardly be surprising and may even be inevitable that the perspective adopted in U.S. news, itself produced by large profit-oriented institutions, is compatible with these interests. However, these dynamics strain the credibility of those who maintain that U.S. news, especially news of science, is fundamentally objective.

The irony remains that this kind of shortsighted approach does not serve the long-term interests of corporate America particularly well. As has already been suggested, one of the few enduring tenets of persuasion research is the conclusion that one-sided messages are ineffective. They are unlikely to be convincing except for audiences of low education who are never likely to hear the other side. In other words, in the modern world (where totalitarian control of communications is increasingly difficult to sustain and literacy is promoted almost everywhere), it is unlikely that anyone can get away with the kind of propagandistic approach to public relations that involves telling only the good news and only from one perspective. Those who want to promote adoption of modern biotechnologies, whether with commercial, scientific, or developmental goals in mind, should welcome debate, even where it involves criticism. Sometimes the outcome may be rejection—the price of democracy, perhaps. But even this worst-case outcome is preferable to building a market on the foundation of a public consensus that is so fragile it subsequently collapses altogether. A world economy dominated by science and technology requires high education and open communication, and this means populations that will demand to be consulted even on complex issues.

NOTES

1. This is all the more remarkable because preexisting patent laws had only been applied to "inventions," not "discoveries," up until that time. It is difficult to argue, on common sense grounds, that identifying the genetic code sequence associated with a particular property or characteristic is an invention rather than a discovery. This issue has become one aspect of the European patent debate.

2. The newspaper data presented in this section draws heavily from material presented in a discussion paper published in May of 1995 by the Texas A&M University Center for Biotechnology Policy and Ethics, later known as the Center for Science and Technology Policy and Ethics, by Asha Krishnakumar, Special Correspondent, *Frontline* magazine (Madras), and Susanna Hornig Priest. Some background material on the GATT was also taken from that unpublished paper. However, the analysis and interpretation are the present author's alone.

3. The U.S. Congress approved the General Agreement on Tariffs and Trade in December 1994.

4. Export of U.S. entertainment products (films and TV programming) represents another form of this "cultural imperialism" (Schiller 1992) that threatens to erase many non-Western values and traditions altogether.

4

U.S. Public Opinion Emerges

In a sense, societal concerns and biotechnology research may be on a collision course, confronting scientists, administrators, and policy makers with difficult decisions, new challenges, and multiple opportunities. . . . Scientists dealing with the public should avoid arrogantly dismissing lay hypotheses; ignoring nonscientific points about biotechnology; and underestimating the intelligence or power of the public.

—Sociologists William B. Lacy, Lawrence R. Busch, and Laura Lacy in their report on "Public Perceptions of Agricultural Biotechnology" (1991)

While the press does not determine public opinion, the public opinion climate provides the context in which information from media sources is interpreted. U.S. public opinion about agricultural biotechnology, as indicated by many poll reports and as suggested by the generally sporadic resistance reported to such events as the introduction of genetically engineered foods, seems on the surface to have remained reasonably positive, especially in comparison to that in other nations around the world. Articles about the medical, as well as the agricultural, uses of biotech have begun to appear routinely in mainstream magazines, where they now tend to be described as foregone conclusions rather than cutting-edge but controversial new technologies. But recent history also suggests that the hegemonic character of our media environment, with its tendency to inhibit the expression of dissent in a "spiral of silence" cycle, has masked a more complex opinion environment. In fact, U.S. public opinion has never been as uniformly positive as superficial attention to news reports might have suggested.

Whether or not a more vigorous and sustained public debate would have created a much different opinion climate in the U.S. in this area—whether more negative or more positive—must remain a matter of speculation, along with the question of whether an important degree of volatility persists or early doubts have largely dissipated. However, the fact remains that U.S. public opinion was probably never as consistently and firmly positive as early reports (including most media accounts) made it out to be, another reflection of mainstreaming effects. In other words, opinion poll data has served primarily to reinforce the appearance of uniformity suggested in other ways by the media. If we define our objective as understanding why the existence of alternative views (outside of delegitimized fringe positions) was not always fully apparent at an earlier stage, rather than why reactions in other societies are so much more "irrational" than the reaction in our own, a more complicated picture emerges that actually makes a good deal more sense.

When the U.S. public responds with vigor to particular perceived threats to public safety, this often seems to come as a shock to stakeholder corporate interests and government officials alike. The result is that the media are routinely blamed for creating what are then easily explained away as exaggerated or unwarranted fears (e.g., as argued in Leahy and Mazur 1980). Given the thoroughly documented existence of agenda-setting effects, there is a grain of truth in this assertion, but only a grain. The media do call our attention to problematic events and circumstances related to food safety, ranging from fast food hamburger poisonings to the alleged carcinogenic properties of pesticides used on fruit. They sometimes do so in an attention-grabbing and perhaps not entirely even-handed way, and their motivation is certainly at least partly rooted in their own commercial interests (that is, they presumably want to grab audiences with these sensationalized accounts in order to make a profit, rather than for purely altruistic reasons). But this is only part of the picture, since media accounts cannot create public opinion out of thin air. For the news media to have much impact, the news must touch on issues that people already believe are important.

ASSESSING PUBLIC RESPONSE

To a large extent, agenda-setting activity represents the mass media's legitimate role, as well as its predominant social effect; we rely on the mass media to survey the environment and identify potential threats (Lasswell 1949), and if they did not do so they would not be doing their job and we would be disappointed. Media institutions are, it is true, much better organized to respond to short-term crises (wars, fires, food contamination, toxic releases) than to longer-term circumstances (environmental deterioration, urban transportation, overpopulation, global warming). But this does not mean the media are responsible for the vigor of the public's response, which has other explanations—notably the dynamics that serve to keep some concerns simmering below the threshold of public visibility in the normal course of events until a tangible drama unfolds that serves to ignite those concerns. A match lit in pervasive dampness is unlikely to start a fire, as many a weekend camper can attest. The volatility of public response to short-term crises is of necessity a reflection of more deeply held values and concerns. In the case of biotechnology, the concerns appear to be widely distributed and quite varied.

Another manifestation of this phenomenon of the masked diversity and volatility of U.S. public opinion with respect to biotechnology and in a number of other areas is the emergence of underground countercurrents that erupt into visibility only when events in the public arena seem to threaten their specialized interests. For example, the U.S. Department of Agriculture was overwhelmed with public objections to its proposal to allow genetically engineered foods to be labeled "organic" if they met the other criteria of this designation, such as having been grown without the use of chemical pesticides.[1] Organic food consumers are not typical of the American public; for one thing, they are a well-organized interest segment, and for another, they are relatively unlikely to accept official interpretations and reassurances at face value. Nevertheless, they are publicly invisible in the normal course of events, though the increasingly common appearance of organic food products on U.S. grocery shelves, even those of the larger chains, suggests that some of these food corporations have discovered this group to be not nearly as tiny a minority as might otherwise have been assumed. The existence of dissident groups like this tends to be masked in the large-scale polls used to gauge public opinion in this country, as well as ignored in major media accounts.

The polls are powerful technology, but they have other problems. They are subject to spiral of silence effects; while the *statistical* reliability of poll data can be determined in a straightforward way, the proportion of people who are actually giving accurate and candid answers generally remains unknown. The lack of accuracy or candor need not be intended as deceptive for poll results to reflect largely unconsidered responses to a particular set of questions asked in a particular order within a particular moment's climate of perceived general opinion. The polls are also not good at measuring the strength or volatility of an opinion, even though such questions are occasionally asked (e.g., "How strongly do you feel about this issue?"). Their failure to distinguish effectively between opinions based on thoughtful analysis and off the cuff reactions makes them poor predictors of likely public responses given additional attention and information, especially in novel and technically complex areas such as biotechnology. The volatility of the public response to issues such as cloning and terminator gene technology is otherwise difficult to fathom. And opinion polls are usually highly instrumental in intent (Davison, Barns, and Schibeci 1997); that is, they are most commonly intended as tools to help guide or manipulate public opinion towards a predetermined end, rather than being designed to create bona fide opportunities for understanding public thinking and incorporating it into public policy.

In this case, of course, the predetermined end is broad acceptance of genetic engineering and other new biotechnologies. This instrumental intent tends to determine (that is, to limit) the type of questions that are asked, in comparison to the type of questions that might be asked in a conscious attempt to highlight diversity of perspective and understand objections rather than to document the degree of apparent acquiescence (which is not at all the same thing as consensus). This is not to say that major polls done by responsible social scientists are in any sense deliberately manipulative, although polls done by less responsible or less experienced persons—especially those sponsored by particular interest groups—can certainly have both this intent and this effect. Rather, the underlying thinking that assumes public opinion to be a monolithic quality best assessed along a unidimensional positive-negative scale itself tends to obscure the existence of alternative points of view. It does so primarily by limiting the questions asked to those that are consistent with this limited understanding of public opinion, but also by limiting the analysis to consideration of the apparent influence of standard demographic categories (age, gender, ethnicity, education, region, and so on) rather than variations in such less tangible but more vital factors such as lifestyle, beliefs, values. In many cases even political or religious preferences are ignored. Contradictory data are taken as evidence of "cross pressures" induced by the existence of conflicting loyalties, rather than as reasonable variations among complex individuals in the interpretation of situations—differences that can be both rational and understandable. Of course good polls give more subtle results, but the tendency to oversimplify and reify is a pervasive one.

Alternatives to the polls exist. Offering opportunities for public comment, such as the procedures incorporated in the promulgation of new federal rules and regulations or in the development of environmental impact statements, generally creates better tools for the purpose of elucidating public thinking for policy making purposes. Unlike opinion polls, these processes do tend to overrepresent the perspectives of special interests, both stakeholders and activists, who of course are those most likely to take the time and make the effort to respond. However, at least they have the potential to tap into alternative points of view in ways poll questions typically do not. Some biotechnology-oriented opinion polls have included religion as an independent variable because of the recognition that religion is likely to be associated with ethical positions on related issues like abortion and fetal tissue use. But there is still little apparent effort to establish a theoretical link between such a variable and patterns of thought and response. Other relevant and meaningful factors such as levels of environmental concern, of faith in science and technology, or of perceptions of conflict among competing economic interests also tend to take a back seat in the analysis.

As a result, meaningful understanding of the relationships among values, beliefs, and attitudes remains elusive. Public opinion is made to seem erratic, arbitrary, and easily manipulated, as though individual members of the public were robots responding to the effects of their own demographic profiles. Policy-makers seem justified in ignoring public opinion under these circumstances. The consensus conference approach taken in a number of other countries, in which smaller numbers of citizens deliberate on the issues facing them instead of responding to close-ended questions, is specifically intended to compensate for the limitations of both polls and open comment opportunities, but this strategy has been employed only rarely in the United States. One of the strongest democracies in the world, we are wary of the public's ability to consider technologies in a reasoned way. Instead, we tend to leave the management of technology to the marketplace and to the technocrats, even though the lone major federal agency charged with evaluating technologies (the U.S. Office of Technology Assessment or OTA, which did the earliest significant work on U.S. public opinion about biotechnology) was abolished in 1995.

Answers to poll questions—and responses to public policy questions and food safety crises—are in fact determined in part by ambient cultural beliefs, which are quite variant in the United States (although certainly patterned), as well as by knowledge and beliefs about genetics. As cultural phenomena, these patterns are best understood holistically rather than as isolated variables; this has been the major contribution of anthropology to social science. The concept of biological heritage is extremely important in American culture (Nelkin and Lindee 1995). Despite America's longstanding romance with science and technology, however, there are other undercurrents of belief many scientists would prefer remained underground. Even though many scientists are themselves religious, strong religious beliefs, on which America also sees itself as having been founded, are widely perceived as being a threat to strongly scientific thinking, a conflict most visible in the controversy over teaching evolution (as opposed to religious creationism) in U.S. public schools. This conflict has reemerged in recent years in a number of states after having been largely quiescent for some decades. It is thus understandable, if regrettable, that some segments of the scientific community appear predisposed to avoiding confrontation with popular sentiment on significant and potentially controversial policy matters. But we are a democracy, and failure to involve the public in policy debates is a risky choice on both ethical and strategic grounds.

Relevant collective social history, which in this case means experience with technology, is also a powerful influence. This includes the series of technological crises and catastrophes to which we have been subject in recent years: Three Mile Island, Love Canal, Chernobyl, Bhopal, the Challenger, and so on. These experiences—along with a smattering of reports of malfeasance in scientific research—have tended to undermine blind public faith in the major institutions of science, technology, government, and the corporate world vis-á-vis their management of technology. It is in no sense an unreasonable or irrational thing to let experience influence trust in this way, but again the influence of such "irrelevant" factors can be used as evidence that public opinion is "irrational." At the same time, the alienation resulting from this kind of distrust puts the public further at odds with the large-institution point of view that the media rely on in reporting developments in biotech.

These kinds of relationships are not obvious from an examination of poll results. Nevertheless, despite all of these limits, and only understood in the context of the bigger picture, the poll data are quite useful for cultivating a deeper understanding of the U.S. public's perspectives on biotechnology. These data need to be examined in the context of lay perspectives on technological risk and levels of scientific literacy, another complex and politically charged issue in itself. Ambient cultural beliefs related to genetics will be explored in the chapter that follows.

U.S. PUBLIC OPINION AND THE POLLS

In 1987, the Office of Technology Assessment (OTA) released their major report on public opinion about biotechnology; the study incorporated some poll data from work done earlier in the 1980s along with their own extensive survey of 1273 respondents done in 1986, making some inferences about trends possible. Unlike some other opinion polls, this study did ask questions about such items as religious beliefs (although only about degree of belief, rather than kind of or adherence to specific beliefs), environmental activism, and faith in science, technology, and relevant institutions, as well as about levels of knowledge and interest with respect to science and science policy. In other words, this poll (done for OTA by the well-known survey firm Louis Harris and Associates) used a very sophisticated approach as opinion-gathering efforts go. Even so, it is often difficult to understand how respondents might have been thinking or the extent to which their responses were informed or thoughtful at this early stage in the public history of biotechnology.

Since biotechnology in the modern sense was at the time so new, it seems reasonable to suspect that many respondents were unfamiliar with the state of the art in this area. However, this seems not to have been entirely the case. According to OTA, about two-thirds of respondents felt they understood the term "genetic engineering" and 85 percent understood the word "gene." While there is some doubt about actual levels of understanding (as judged by ability to explain, which may or may not be the most reasonable method of assessing understanding), OTA estimated that about half had "a good general sense" of what genetic engineering was all about; more than three-fourths of the respondents reported having heard or read at least a little bit about genetic engineering (1987, 47).

The study also found that concern about science policy was present "across all demographic subpopulations" (1987, 19). It seems as though awareness, if not necessarily knowledge, was already widely distributed.

It is on the basis of this extensive report that the U.S. public was generally reported to have been optimistic about biotechnology, an impression that was carried by the early trade press and media reports and that hardly wavered in subsequent public accounts. This impression fit into the corporate campaign for agricultural biotech very nicely. But in reality, the report concluded that public enthusiasm for genetic engineering had already declined in comparison to data from 1982 (1987, 50). The proportion of people who thought that genetic engineering would improve the quality of life "a lot" went down from 32 percent in 1982 to 18 percent in 1986.

Despite a general level of technological optimism that seemed to OTA to have persisted throughout the environmental and technological disasters of the 1980s, more people wanted biotechnology more aggressively regulated than had been the case a decade earlier. Fully 77 percent agreed with the statement that genetically altered cells and microbes called for "strict regulations" (1987, 81). And 61 percent agreed that we have seen only "the tip of the iceberg" with regard to biotechnology's risks (1987, 29), even while 59 percent agreed that most technology-related risks that cause people to worry never happen. Finally, 46 percent agreed that "We have no business meddling with nature" (1987, 81). While not a majority opinion, this is hardly a reflection of consensus either. The report also revealed "pervasive" awareness of environmental issues not limited to those who were especially attuned to science-related issues (1987, 37), possibly an unrecognized harbinger of later concerns. Yet, perhaps because attitudes toward specific applications were usually more positive than abstract opinions, the report was presented and widely interpreted as reflecting resilient American optimism.

Roughly equivalent public opinion data from Europe are available in the Eurobarometer series, results of which extend through 1996. During 1996, averaged across the seventeen participating European countries, this survey found 22 percent of respondents believed that biotechnology or genetic engineering[2] was likely to make their "way of life" worse. *This is exactly the same percentage of U.S. respondents in the 1986 OTA survey who thought that genetic engineering would have a negative effect on their quality of life.* Two caveats are necessary. First, question wording used because respondents to the U.S. OTA survey were not given a "no effect" option but were asked to choose between better or worse outcomes. As a result, the data are not exactly comparable. Given more options, only 44 percent of Europeans (but 66 percent of Americans) said they anticipated a positive impact. Secondly, among European respondents asked this question about genetic engineering specifically, the percent envisioning a negative impact was somewhat higher (27 percent), and the percent envisioning a positive one somewhat less (39 percent). The Europeans are more cautious, but the contrast is not complete. The U.S. public also included significant numbers of people who were not persuaded biotechnology was good.

The OTA report can reasonably be taken as reflecting relative optimism among U.S. consumers. Nevertheless, it also seems clear that there was nearly as much pessimism lying beneath the surface in the United States of the 1980s as there was a decade later in the Europe of the 1990s, after extended public debate there, and that the focus on the optimistic segment of the U.S. population at a time when little publicity about these developments had as yet taken place may have been quite misleading. Other early data from a variety of sources also showed substantial proportions of U.S. consumers were concerned to some degree about biotechnology-based products in the food supply (USOTA 1992). Subsequent surveys (reviewed in Hoban and Katic 1998) continued to show most—but not all—consumers would purchase fruits and vegetables produced using biotechnology, with about one-fourth to one-third unlikely to purchase them. While the data for 1997 suggested better-educated respondents were somewhat more likely to find these products acceptable, all demographic variables combined accounted for less than 4 percent of the variance in acceptability. That is, the variations in biotech-related opinion did not appear to follow patterns defined by membership in standard demographic groups.

Science and Engineering Indicators 1998, the most recent in a long series of major National Science Board reports published by the U.S. National Science Foundation that addresses public understanding of science along with many other broad trends considered vital to the interests of the scientific community, contains more recent data on public attitudes toward biotechnology, in particular genetic engineering. According to this report, which includes data from 1985, 1990, 1995, and 1997, just 42 percent of Americans now believe that the benefits of genetic engineering will outweigh its harmful effects, while 36 percent believe that the harm will outweigh the good (National Science Board 1998, 7–14). Both of these categories—but especially the first, more positive one—have actually shrunk slightly since 1985, when the respective figures were 49 percent and 39 percent, with the "about equal" category having taken up much of the slack (A-407). In other words, these data also show a significant percentage of the U.S. population, although not a statistical majority, expressing reservations about genetic engineering, even in the early years.

While expanded discussion in the media and elsewhere may have increased the proportion who were undecided or felt that harm and benefits were roughly equal, as the report speculates, the underlying pattern is more one of stability than change. In other words, a significant proportion of the U.S. population has always had reservations about genetic engineering and these have not gone away. This decline was evident despite persistently high levels of optimism about the benefits of scientific research generally. Nor can they reasonably be attributed to the influence of the media or other information sources, even though some erosion in optimism might well be a result of respondents hearing about issues that were not immediately obvious to them fifteen years ago.

Given the well-established agenda-setting effect of the media, a media agenda that had been excessively focused on problems with genetic engineering over the years would presumably have generated high percentages of the public attuned to relevant issues. This is not the case; the proportion of the U.S. public demonstrated to be attentive to related issue areas is generally modest, including the proportion attentive to new inventions and technologies (only 9 percent), to science and technology policy (14 percent), to medical discoveries (19 percent), and to environmental issues (12 percent) (National Science Board 1998, A-392). These data do present a profile of the U.S. public as relatively complacent about such questions, but this is not the same as being actively supportive or accepting of biotechnology in agriculture and not necessarily predictive of later reactions.

We now have four reasonable answers to the question of why the U.S. public did not object to the introduction of bioengineered foods while Europeans did, none of which entail reliance on a hypothesized media-based Frankenfood hysteria effect taking place in Europe. First, *U.S. consumers did question these technologies*, although perhaps not in quite such large numbers, and they continue to do so; early evidence of unvoiced dissent and a downward trend in optimism were widely ignored by interests bent on introducing bioengineered products into the U.S. food supply in short order. Second, by overrepresenting the large-institution point of view and the ostensibly monolithic character of U.S. public opinion, *media accounts probably helped to suppress the visibility of what dissent existed*. Third, again as a result of the early domination of this point of view in the U.S. press, as well as (in circular fashion) the illusion of public consensus thus produced, *only positive messages were likely to be legitimized* in U.S. news accounts. This is true even though individual journalists and editors probably strove to produce what they saw as balanced stories. And finally, *spiral of silence effects very likely mitigated against public expression of differences*, with the exception of objections voiced by a handful of dissenters subject to being perceived, presented and dismissed as extremists.

SCIENCE LITERACY AND BIOTECHNOLOGY

The 1986 OTA report had noted that levels of education and the belief that genetic engineering would improve the quality of life were largely unrelated (1986, 51). Although self-reported level of science understanding was associated with a greater likelihood of this optimistic belief, it is difficult to distinguish self-reported understanding in the general population from enthusiasm for science. Among agricultural science faculty, on the other hand, self-reported knowledge has very little relationship to positions on policy issues (Priest and Gillespie 1999). Some researchers have identified a relationship between education and acceptance of food biotech (Hoban and Katic 1998), but the relationship appears weak. Others have pointed out that, at least in the United Kingdom, increased levels of understanding may lead to more, rather than less, concern over aspects of science seen as morally problematic (Evans and Durant 1995). Nevertheless, the scientific community and biotech's corporate promoters seem determined to interpret objections to biotechnology to a misunderstanding of biotechnology. Levels of science literacy are implicated in this debate; indeed, they have become something of a weapon in policy battles.

Science literacy is a major theme of the relevant sections of the 1998 NSF report on science and engineering indicators; by setting the discussion of public assessments of various technologies, including genetic engineering, in this context, there is some implication that adequate levels of education could eliminate dissent. However, the report itself does not support such an interpretation, even though this may have been an implicit hypothesis of its designers. Rather, increases in uncertainty or perceptions of an equal balance between benefits and harms were seen in high school graduates as well as college graduates (National Science Board 1998, 7–14). Even for those with a bachelor's degree or higher, 24 percent continued to believe in 1997 that the harms probably outweighed the benefits (in comparison to 37 percent of high school graduates and 44 percent of those without a high school degree), as did 27 percent of those defined as members of the "attentive public" for science and technology and 30 percent of the "attentive public" for medical research (7–22). Concerns are clearly not limited to those without much knowledge of science.

Regardless of the impact on opinions and attitudes, do Americans really understand genetic engineering and other new biotechnologies? The National Science Board report concluded that only 22 percent of the population could give a reasonable definition of DNA (1998, A-394). Questions about how these items were evaluated are raised, however, by the fact that only 11 percent of the population could give a reasonable definition of a molecule, only 13 percent could define the Internet, and only 11 percent could define radiation. It is counterintuitive that twice the number of people could define DNA as could define "molecule" or "radiation," especially since DNA is most commonly represented *as* a molecule (the familiar ladder structure). In fact, a slim majority (51 percent) of those *with graduate or professional degrees* failed the DNA question, as did almost half (44 percent) of those defined as having a high level of science/ mathematics education (nine or more high school and college courses) and nearly two-thirds (66 percent) of the "attentive public" for science (National Science Board 1998, A-394). Well over half (57 percent) of those surveyed had access to computers at work, at home, or both (A-413), yet at least three-fourths of these must not understand what the Internet is if the statistic cited above is correct. Similarly, only 27 percent of American adults, 54 percent of those with graduate or professional degrees, and 55 percent of those with a high level of science and mathematics education seem to understand what scientific inquiry is all about (A-395), at least by this assessment.

It does not seem wise to take the answers to these questions entirely at face value. Data like these on science literacy are excellent arguments for improving science education, which is one of the reasons they are gathered. Unfortunately, on the surface they might also seem to be good arguments for either ignoring public opinion on scientific and technological controversies or treating it as something to be educated away. But it does not follow that because someone cannot give an acceptable definition of

DNA that they could or should have no voice in arguments about biotechnology. In a world in which many, if not most, public policy decisions have at least some connection to science, technology, medicine, or agriculture, to exclude nonexpert opinions would be enormously poor democracy. Rather, these data would appear to be good support for encouraging the broadest possible public debate, if only to allow the scientific background to these controversies to become better known.

At the same time, while the performance of U.S. respondents on tests of general science literacy does not seem impressive, for whatever reason, the United States still scores well on such tests in comparison to most other countries of the world. In each year for which data are available (1990, 1995, and 1997), the United States scored at or near the top of the range in scientific construct understanding as measured for thirteen other countries in 1991 or 1992. However, the mean scores for Denmark, the Netherlands, Great Britain, France, and Germany (ranging from fifty-one to fifty-five) were very close to the U.S. scores (fifty-four in 1990 and fifty-five in both 1995 and 1997) (National Science Board 1998, A-396). Objections to genetic engineering have been raised in all of these countries, additional evidence that reservations about biotechnology cannot be reduced to an educational deficit issue.

Finally, the United States scores the lowest of all of these countries on a measure of belief in "scientific reservation" while scoring about the same as twelve of the other thirteen on a measure of "scientific promise" (National Science Board 1998, A-400). Japan was the lone exception, scoring fifty-five on the index on which all of the other countries included ranged from sixty-four to seventy-five. This is strong justification for arguing American optimism in this area persists. But the "scientific reservation" index was composed of three items: excessive dependence on science as opposed to faith,[3] importance of knowing science in daily life, and pace of science-induced change in way of life. Just as many U.S. respondents agree as disagree (47 percent in each case) that we are too dependent on science instead of faith (A-398). However, this result is obscured in the composite index because 85 percent do not agree that science is "not important" in their daily lives, and 61 percent do not agree that science makes change occur too quickly. In general, only 12 percent of Americans believe the "harmful results" of science outweigh its benefits (A-398). In short, many U.S. citizens may or may not be comfortable with the role of science in their society, even though they use it in their daily lives, are not disturbed by the pace of change they attribute to it, and do not feel it is a harmful influence overall.

Americans are generally positive about science, and while they may not understand it as well as some would hope, score at least as well as the citizens of most other developed countries on tests of science literacy. Yet many of them do have reservations, both about science in general and about genetic engineering specifically, just as their European counterparts do. There is no clear evidence that these reservations are reducible to misunderstandings or poor science education, even though it is difficult to argue with the assertion that science education could and should be improved. This conclusion is consistent with what is known about how judgements about technological riskiness are, in general, derived: perceived risk is not a straightforward function of probability of harm, but neither is it always a function of educational deficits. A broad range of sociopolitical and psychological factors figure in the lay public's perceptions of all technology-related risks, not just those associated with biotechnology.

PERCEPTIONS OF TECHNOLOGICAL RISK

Extensive research and other scholarship over the past several decades have demonstrated that the nonscientific, or lay, public understands technological risk in different, although not necessarily erroneous, ways from the manner in which the scientific community understands technological risk.[4] To nonscientists, the risks of a technology are not just a straightforward matter of the probability of a harm (or "hazard," to use the language of risk analysis). The kind of harm envisioned—whether known or unknown, familiar or novel, common or "dread"—is also important (Slovic, Fischoff, and Lichtenstein 1979). Such things as the perceived adequacy of regulatory mechanisms, the behavior of relevant political processes, and the ethics and intentions of those responsible for creating and managing the risk are also important (Hornig 1992a; Priest 1995). These elements heavily influence the interpretation of new information received through the media and other sources.

Gender differences in responses to technological risk may actually be attributable to subtle differences in values and attitudes held by men and women in our culture (Hornig 1992b), differences that persist despite expanded professional opportunities for women. These differences are a valuable illustration of the fact that disputes about risk perception represent something more than a question of accurate versus inaccurate understanding of data. We do not need to see the higher levels of perceived technological risk that are characteristic of women as being a function of those women being less well educated or less likely to understand data-based judgements. Rather, such differences as a higher valuation of social and familial elements may cause women to be more concerned about technologies seen as having the potential for disrupting the social fabric. This is not an issue of rational versus irrational responses to risk; it is a difference in judgement based on differences in values and priorities that presumably result from differential socialization. For this purpose men and women can be usefully thought of as members of different subcultures.

More generally, while nonexperts can and sometimes do misunderstand the nature of risks and misjudge their magnitude (as do the experts, on occasion), it is common for scientists and nonscientists who disagree about risks to be basing their judgements on different criteria. The nonscientist includes a broader range of issues that the scientist is trained to exclude. Can this official really be believed? Who stands to gain or lose by defining this risk one way or another? Would a catastrophe associated with this technology destroy a way of life, or (less dramatically) impact its quality? What happened the last time a similar technology was deployed? Lay publics tend to ask these kinds of questions along with narrower questions about the technical probability of a harmful occurrence, while scientists want to keep the two separate. We might productively describe nonscientists as working with an "expanded vocabulary of risk" that may actually be a better model of important real-world considerations than the narrower conceptualization that scientists and many risk managers have been trained to employ.

None of this is to say that scientists are wrong and nonscientists are right. But it is important not to assume that lay judgements about risks are necessarily irrational, either. Basing judgements on so-called emotional criteria such as trust, fear, and perceived threat is a perfectly rational response, and quite probably an adaptive one from an evolutionary perspective. The answers to the question quoted above from the 1998 *Science Indicators* survey does indicate some discomfort with substituting science for "faith" (presumably religious faith); however, data from a 1996 survey also included in that report indicate that the general public has more confidence in the scientific community than they do in organized religion (with 39 percent and 25 percent, respectively, having "great confidence" in the leadership of these institutions), with confidence in major companies not very far behind the latter at 23 percent (National Science Board 1998, A-403). As the OTA report had concluded a decade earlier, "moral objections to genetic engineering of plants or animals" is based on "a broad range of beliefs, concerns, and fears that go well beyond religious issues" (1986, 58).

Fears of confrontation over the status of evolutionary theory or other prima facie irreconcilable differences between science and some branches of U.S. religions should not be used as a justification for ignoring public opinion on these issues, nor can most public objections be dismissed as functions of fundamentalist perspectives that reject science. The same seems to be true in Europe, where only 3.3 percent of the 1997 Eurobarometer respondents indicated any trust in religious organizations to tell them the truth about biotechnology (Durant, Bauer, and Gaskell, eds. 1998, 267). While over one-third of Europeans (36 percent) tend to agree that religious organizations should "have their say" in how biotech should be regulated, almost half (49 percent) do not (263)—and even having "a say," of course, is not the same as dictating policy. Religion is unlikely to be behind objections to biotechnology in either Europe or the United States.

However, cultural beliefs about genetics are another matter. While opposition to biotechnology cannot be reduced to narrow religious objections, the interpretation of messages about all risks and benefits takes place in a broader cultural context that, for the reasons discussed in the introduction to this chapter, are not well reflected in public opinion statistics. These views are not always rational, well informed, or consistent with available evidence, any more than they are always irrational or based on misinformation. Long term, the media undoubtedly help shape these cultural beliefs, but they do not determine them. These cultural elements and their potential importance are the subject of the next chapter.

NOTES

1. One argument put forth by biotech's promoters has been that engineered foods can be created which can be produced in more environmentally friendly ways. So far, however, there have been more apparent failures than recognized successes along these lines; for example, pesticide-resistant crops have recently been blamed for harming monarch butterfly populations. It may be that failures of this type are more likely to be considered news than are successes.

2. Question wording alternated systematically between these two items.

3. The wording choice here is probably unfortunate, a significant lapse in an otherwise highly professional survey project. In this kind of "double-barreled" question, it is impossible to tell whether respondents are really commenting on dependence on science per se or are more concerned with the implication that science is replacing (religious) faith. While the United States is arguably a highly religious country, at least in terms of its historical legacy, many people seem to hold scientific and religious beliefs simultaneously without any apparent internal conflict. For this reason, results from a survey question that appears to require them to choose between the two are especially difficult to interpret.

4. Some scholars (e.g., Bradbury 1989) argue that use of the term "risk perception" is value-laden, especially when applied to the thinking of nonscientists, because it is seen to assume that there is a single "correct" conclusion about any given risk against which lay perceptions should be judged (and will be found wanting). The term as used here should not be taken to involve that kind of an assumption.

5

U.S. Culture and the New Genetics

I am pro-genetic engineering and did not realize that many people. . .would be against such "tinkering" . . . . People have raised questions about who is going to regulate such activities and who should be able to "play God."

—College student responding to question (Priest and Taylor 1995)

Public opinion is not formed from media messages, but it is in important ways a product of the interaction of culture with those messages. And culture, in turn, is embedded in popular media products, both news and entertainment. This process helps to reinforce cultural values and beliefs, although it does not create them out of nothing. As we have seen, access to media is not equally distributed; the general cultural perspectives, as well as the specific opinions, of large-institution stakeholders are overrepresented. Even popular culture products such as MTV, romance novels, tabloids, and talk shows reflect a worldview consistent with the status quo distribution of power. This is especially so because of the influence of advertising interests on commercial entertainment in the United States. What better promotion could materialism hope for than a show like *The Price is Right* or *Wheel of Fortune*? Other popular culture messages may be less clearly tied to political or economic interests, but are still influential and important. What could be a more effective arbitrator and enforcer of changing social norms than the soap operas and talk shows that identify and publicly chastise offenders and set our social issues agenda? Media do not have many direct or "magic bullet" effects, but they do incorporate cultural values and other socially significant messages.

Advocates for women and for ethnic minorities quite rightly concern themselves with how these groups are portrayed on entertainment television. So do doctors, lawyers, and engineers. Far from trivial, such images are important contributors to the social construction of reality—they provide us with concrete images of appropriate social roles and of settings ranging from operating theaters to judges' chambers to battlegrounds with which we will rarely, if ever, have direct personal experience. These effects need not be exclusively negative. The world may be a better and more peaceful place because we have clearer images of the humanity of peoples around the globe, as well as of the impacts of wars, famine, and natural disasters in distant places. And we certainly know much more about science than previous generations, not only because of the proliferation of science-oriented magazines that appeal to specialized audiences but also because of mass-appeal television and the integration of scientific and medical news with news of other types.

Culture is not limited to that which is embedded in the media but is a much more pervasive and persistent influence. Culture is often thought of in broad outline as a way of life or a worldview learned as a member of a particular social group, especially the group into which we are born. Culture provides the context for perceiving, understanding, and planning everyday activities. It includes language. One useful way of thinking about culture is that it is the knowledge that enables the members of a social group to get through their everyday lives. This kind of knowledge also enables those people who live in societies with developed mass media to interpret new media messages. To do this, we need to know not only the language itself but an entire system of formats and symbols—from the structure of a television commercial to the meaning of the American flag; we also need to know who the "good guys" and "bad guys" are and what are considered desirable or undesirable outcomes (social values). Cultural heritage, our legacy of learned and passed-down information, provides the store of knowledge and beliefs on which such judgements are based.

Of course, culture can also be a barrier to change, as when religious beliefs or social practices inhibit people's willingness to accept medical, nutritional, or agricultural advances. Even in cases where medical intervention could save a child's life, legal and medical practice in the United States leans very much in the direction of respecting cultural (or subcultural) differences and preserving the right of individual choice wherever personal preferences and scientific or medical practice are in conflict, especially where the issue involves religious beliefs. In such instances, we tend to work through persuasion and education rather than force. But conflicts of this type tend to be the exception rather than the rule. In general, the fundamental faith in science that is characteristic of U.S. culture and our equation of scientific discovery with social progress are cultural features that normally predispose us to accept new technological developments, including agricultural biotechnology, even though our general optimism about technology has been challenged in recent years by a series of catastrophes and environmental crises. Other strands of our culturally derived knowledge and belief are likely to lead us in the opposite direction, however, sometimes undermining attempts to "sell" genetic engineering and other products of biotechnology on the basis of their scientific promise. These are likely to include the belief that certain biological distinctions are inviolate, that biology largely determines fate, and that genetic heritage should not be manipulated, as well as faith in a democratic ideal that still champions the rights of the individual and distrusts the motives of large institutions, whether corporate or governmental, when these seem at odds with individual freedoms.

Such beliefs are neither wrong nor right, nor are they very often readily classified as either leftist or conservative. Many of these currents tend to be found at all points on the American political spectrum, although in somewhat different forms. But they do go a long way toward explaining public response in the United States to more recent forms of biotechnology, from cloning to terminator technology, after years of apparent complacence. Public response to such developments is more likely to be based on deeply held cultural beliefs and values rather than quixotic opinions or limited comprehension of the relevant scientific ideas. This means that public objections would not be easy to extinguish even if the complex ethical barriers to this kind of manipulation could be set aside.

CULTURE AND BIOLOGICAL BOUNDARIES

One important task that every sentient being must accomplish is organizing the world of sensory perception. For human beings, culture tells us how to do this. Linguistic labels—words for things—that we learn as members of a particular culture provide us with a set of categories that is our most important tool for this purpose. Whether this means of classifying phenomena is or is not entirely limited to human beings; the nature of this interaction between raw, biologically "wired" perceptions and superimposed cultural interpretations; and the extent to which linguistic systems, once learned, serve as active constraints on our thinking are all issues about which professional opinion among anthropologists and psychologists remains somewhat divided. But there is little doubt that learned, culturally determined, linguistic categories help people to organize the world of perception.

A key characteristic of linguistic systems seems to be that they incorporate distinctions or boundaries that are sometimes so fundamental they can be considered sacred. In our culture, for instance, gender is one such category. People who are not clearly classifiable as either men or women confuse us. If we meet such a person on the street, or see one on television, we try to reconcile what we observe with one category or the other, sometimes provoking a perceptual shifting back and forth as though we were studying a psychologist's figure-ground puzzle in which first one shape and then another is perceived as dominant. We find comedy routines involving recognizable entertainers dressing as the opposite sex hysterical; when an occasional drama involves an actor actually "passing" as a character of the opposite sex, we are astounded. These are not necessarily so much homophobic responses as they are illustrations of the fundamental character, for us, of the gender distinction. This distinction is one of a series of such distinctions that we treat as biological absolutes.

Another such sacred distinction in our culture is that between the living and the dead. At this point Frankenstein images have become a part of our culture, along with vampires and zombies, but the underlying fascination is not a product of Mary Shelley's pioneering science fiction story. Rather, all of these mythical beings represent longstanding popular concerns with the sacredness and impermeability of this line between the living and the dead and the sanctity of related natural processes. Almost nothing is more terrifying to many of us than the prospect of being buried alive, a predilection some trace to medieval days when this might have been a relatively more common mistake. These concerns are certainly not specific responses to recent biotechnology; they are pervasive and persistent cultural themes that cannot be reduced to questions of scientific knowledge and ignorance.

A willingness to be entertained by horror movies about ghoulish creatures who come back to life (the ever-popular classic *Night of the Living*

Dead comes to mind, along with the more recent and partially tongue-in-cheek *Chuckie* movies about an evil doll that cannot be killed) has little or nothing to do with a belief that such a sequence of events is actually possible. The same applies to being entertained by a blockbuster movie like *Jurassic Park,* in which modern biotechnology brings long dead species of ancient dinosaurs back to life from preserved DNA. We are often repelled and simultaneously fascinated by creatures that cross the line from death back to life; this is a large part of the attraction of these stories. Social policies become controversial when they involve disputes over this boundary: when does a fetus gain consciousness, or when does "brain death" occur? Just where is the boundary between the living and the dead? It is worth mentioning here that not all other cultures maintain this line so rigidly; in many societies, ghosts walk more freely beside the living. Popular culture products that appear in our own society—sci-fi and horror movies, for example—reflect our own cultural preoccupations. These are not necessarily universal, and their appearance in popular media products is not evidence that they are actually products *of* those media.

The line between the living and the dead is not the only such sacred boundary crossed by genetic engineering. The distinctions between plants and animals and especially that between animals and people are also inviolate. Exceptions—things that are neither living nor dead, neither plant nor animal, neither animal nor human—are usually disturbing. It seems especially disturbing to think of injecting human genes into animals, perhaps because we imagine that this will create "smart" animals or animals with human-like consciousness in other respects, creatures that we won't quite know how to categorize without violating this sacred human-animal distinction. What kinds of consideration—what uniquely "human" rights—would such a creature be due? Our reverence for the individual human being demands that this issue be considered. Yet on a day-to-day basis, the fact that we regularly slaughter and consume animals that may "naturally" have human-like consciousness is not a source of much concern except among a few animal welfare enthusiasts, whose arguments we generally marginalize. Rather, it is the genetic mixing that involves blurring of the animal-human boundary, not killing or eating per se, that seems to cause us to lose sleep.

Crossing this boundary between animals and people has also been the subject of popular culture portrayals, as in the 1996 Marlon Brando film, *The Island of Dr. Moreau*, about the creation and demise of an entire small kingdom populated by illicit genetic crosses that combine characteristics—good and bad—of animals and human beings. While the film is based on an old H. G. Wells' story, the DNA splicing technology that the new *Moreau* employs is much more up to date, part of a still-emerging genre of genetic horror stories. And again, it is not necessary to believe that such feats as this and other fictional themes are literally possible to be fascinated by the proposition, and that fascination in itself is a reflection of cultural influences. These reactions presumably spring from a special reverence for human life, a value with which it would be very difficult to quarrel. But here also is to be found the roots of many public concerns about research involving fetal tissues, animal-to-human transplantation, and to some degree all forms of genetic engineering, including agricultural applications. Whether we think of them as natural biological boundaries or something God-given, we do not seem to be comfortable tinkering with species boundaries—especially when the tinkering involves human genes.

The animate-inanimate distinction, the stuff of scores of science fiction films about robots and cyborgs, is a very similar one but actually seems more permeable a boundary than the one between animals and humans. As science fiction movies like *Blade Runner* and a dozen similar ones pose the question, what rights would accrue to artificially created robot slaves with human consciousness? Isn't it the consciousness itself, rather than its nonbiological origins, that is critical? We do not seem to be certain of the answer to this question one way or the other, but we root for Mr. Data from television's *Star Trek: The Next Generation* in his quest to understand and internalize human emotions and to successfully manage subtle and challenging relationships with actual human beings. It does not seem to matter to our sense of identification that Data is a mechanical, rather than a genetic, creation. He is "only" a machine that aspires to be human, but he has human-like consciousness and intentions, and that seems to be what matters.

GENETICS AS NATURE AND FATE

Given the centrality of distinctions between human and nonhuman beings and the special role ascribed to genetic heritage in our culture, we should think more carefully about the idea of genetic fate. Popular thinking about whether traits such as "goodness" or criminal tendencies are inherited or acquired has important public policy consequences. Popular media accounts of identical twins raised apart but following parallel paths through life, and of adults adopted as children striking immediate rapport with recently rediscovered biological families, are all telling related stories, stories that resonate with the belief that genetics is everything. Our culture, as Nelkin and Lindee (1995) have pointed out, is obsessed with biological heritage. We really have very little doubt that in actuality "blood is thicker than water."

Anti-abortion rhetoric argues that once conceived, an embryo has an inalienable right to exist and to be nurtured comparable to that of any other human being, regardless of any defects or the wishes of its progenitors. The message is not only that human life begins at conception, but that once the genetic dice are rolled, it is too late for second-guessing. Needless to say, this argument is hardly universal. Not only is abortion still legal in the United States (and this is apparently supported by a majority of the public, although poll results vary depending on question wording, circumstances, and other factors), but societies around the world have practiced both infanticide and euthanasia consistently throughout human history, and many contemporary nations seem to accept abortion more readily than we do by far. Yet the premise on which abortion resistance in the United States is based extends beyond the arguments of pro-life conservatives. Liberal rhetoric about valuing diversity springs from a very similar root, the assertion that the human individual is sacred and that all human lives are inherently worthy, if not literally equal. This is hardly a value very many of us would question, yet it seems at the same time to often divide us. Those in our society who emphasize the "rights of the unborn" and those who champion the autonomy of people at other stages of life to make their own decisions are not typically the same people, but clearly both groups are operating within the same cultural tradition.

It seems to be primarily in situations in which the principle of individual choice and freedom (an absolute right accorded to all competent adults in our society) seems to compete with the mandate to preserve and value all human life (at whatever stage of development or decay) that conflicts surrounding this value tend to arise, and occur not only where abortion is contemplated but also when the question of the "right" to die is raised. We struggle with questions of whether life should be artificially prolonged at its conclusion, and we generally abhor the ideas of eugenics and euthanasia.[1] The controversies primarily have to do with which interventions, and under what circumstances, exceptions to this natural life cycle are tolerable. But while different factions argue for or against intervening at various points in the process, there can be little doubt that reverence for the natural cycle of human birth and death is a widespread cultural value in the United States. Genetic heritage is the inalienable "given" that underlies this sacred cycle and is disturbed only at our peril. Germ line (inheritable) alteration controversies in medical genetics may spring in large part from this foundation, although they are usually rationalized in terms of unpredictable future consequences. We alter the human gene pool every time we intervene to save the life of a child born with an otherwise fatal inheritable defect, enabling that person to grow up and reproduce, though we rarely if ever question such practices. Yet we find germ line genetic alterations, even in the name of therapy, problematic and difficult to justify.

In this context, the presumption that character, as well as personality, is largely hereditary is not just a mistaken popularization of the recent trend in academic psychology toward biochemical explanations; it is also a belief that seems to have deeper roots in American culture. Such beliefs tend to overstate the relationship between genes and their expression (Hubbard 1995; Hubbard and Wald 1993) and, more importantly, invite us to overlook the social and economic explanations of contemporary social problems (Nelkin 1995). They are in this sense a re-embodiment of earlier cultural beliefs in the primacy of fate as a determinant of human outcomes, with fate now understood in quasi-scientific rather than theological terms as a function of the random genetic combination that is put into place at conception. We see everything that follows as springing from these influences; even though we also tend to hold the individual responsible for his or her subsequence choices, we tend not to seek causes outside of the individual for many pervasive social ills.

While perhaps not as directly relevant to biotechnology in agriculture as to issues of human genetic testing and gene therapy and to more general questions of social policy, such values and beliefs are still part of the backdrop that frames all genetics-related issues, arguably making them more sensitive than they otherwise would be. And the apparently ubiquitous belief that character is hereditary serves to illustrate the persistence of such thinking largely independent of media accounts. Concerns about free will versus determined fate, once confined to the realm of the theological, take on new meanings in the age of the new genetics. Altering genetic structures is equated with playing God. In a culture in which God is conceived of as all-powerful and omniscient, playing God means changing destiny. And destiny is in our genes.

Environmental concerns are sometimes closely analogous in terms of an emphasis on destiny, though this is rarely phrased in religious terms. In other words, reverence for the environment seems to be an extension of earlier modes of thought, rather than representing a radical break from traditional American values; it is old wine in new bottles. We speak of the law of nature and its wisdom, as though the outcomes of natural processes were always and necessarily a matter of just destiny. Evolutionary theory seems to support this idea by asserting that superior characteristics and species are those that have been preserved. We have learned the hard way that to interfere with ecological and evolutionary processes is dangerous. That changing one element of an ecosystem can indeed have unpredictable consequences has been widely recognized since the 1962 publication of Rachel Carson's *Silent Spring* exposed the impact of the pesticide DDT on eggshells, with profound consequences for the population of predatory birds such as the California osprey. Nevertheless, it seems that in some ways a reverence for ecological integrity has become a modern religious tenet. And once again, consistent with our cultural and philosophical traditions, it is a tenet that cautions strongly against excessive human interference in natural cyclic processes—against any attempt to "fool Mother Nature." That ecosystems change and species become extinct *naturally* does not especially concern us; that they should do so as a result of *artificial* human activity that interferes with a natural process does concern us greatly.

CULTURE, MEDIA, AND CLASS

The assumption that such things as character, as well as personality, are hereditary, that genes determine who will be successful and who will become a criminal, is probably not a media phenomenon but has much deeper cultural roots. Some evidence of this comes from a small pilot study of twenty-eight undergraduate college students conducted in 1995 (Priest and Taylor). Conventional mass media—newspapers, magazines, and television—were rated most highly as sources of genetic information by this group, as was formal education. As a whole this group did not believe that successfulness or "criminal behavior" were genetically determined; they were on average approximately neutral about whether personality, intelligence, or disabilities were likely to be inherited. But the data also revealed positive correlations between reliance on "family" as an information source and belief in the inherited nature of successfulness (correlation coefficient of .42; not statistically significant) and of "criminal behavior" (correlation coefficient of .51; statistically significant at the .01 level). In other words, it looks as though such beliefs are more likely to be passed on interpersonally rather than through the media.

News media aimed at different groups of people probably vary in terms of their portrayal of genetic influence as a matter of influence or determination. To explore this question, we compared the content of elite and more popular newspapers for a three-month period extending from January 1 through March 31, 1995, using the National Newspaper Database Index or NNDI for elite coverage and the Newsbank index for a broader, nonelite, cross-section of daily papers. We used these two sources to select twenty-seven articles (sixteen from the NNDI and eleven from Newsbank) that dealt most explicitly with the subject of human genetics, excluding for our purposes those more concerned with the business side of the equation. The content of the NNDI and Newsbank articles appeared roughly comparable in terms of subject coverage. We classified each paragraph within an article as to its predominant source and whether it was primarily about specific people or primarily about science or medicine in general. We classified each article as to the percentage of words implying certainty in relation to the total number of words implying either certainty or uncertainty. Words like cause, responsible, resulting, engineered, induced, known, determine, and so on were classified as implying certainty; words like contribute, suggest, likely, complexity, believe, seems to, and so on were classified as implying uncertainty. And, finally, we also subjectively rated each article as to its adherence to a narrative storyline versus standard "objective" journalism, that is, a more informational style of reporting.

Our results showed that the nonelite papers, that is, the Newsbank material, contained a significantly higher proportion of words implying certainty than did the material from the elite NNDI papers ($P = .05$). The elite media contained fewer paragraphs about science as opposed to people and less background information ($P = .003$ in each case), as well as fewer nonexpert sources (P = .04). The use of a narrative as opposed to an informational style was not much different for the Newsbank versus the NNDI material. In other words, the more popular or nonelite papers published less people-oriented material and included more scientific explanation, and even though they used about the same style of reporting as the elite papers, they were also more inclined to present genetics-related topics using more deterministic language.

Whatever the explanation, the less frequent use of language implying certainty in the elite media accounts is consistent with survey data reported in the previous chapter that associates greater education with greater, rather than fewer, questions and reservations about modern biotechnology. And this result is also consistent with the pilot survey results described above, if we assume that those more dependent on "family" information sources are likely to be consumers of popular, rather than elite, news media products—that is, of those media products most likely to use the language of certainty to describe genetic issues.

On the one hand, this style of thinking—the association of mass U.S. culture with a noncontroversial, deterministic, and matter-of-fact conception of both genetic science and the role of genetic factors—could help explain why many segments of the U.S. public were fairly complacent about biotechnology throughout much of the decade of the nineties. Further research would be needed to confirm this, but biotech may have been both presented and seen very much as scientific progress-as-usual by some segments, in part because genetics is understood fatalistically in the first place and perhaps also because of a less inquiring approach to science in general. At any rate, for many people, no reasons for resistance were encountered, and no alarm bells rang.

On the other hand, it also seems likely that a cultural tendency to consider genetic heritage—along with distinctions between animals and people, between the living and the dead, and between the animate and the inanimate—to be sacred may have predisposed us to the kind of public outcry that met later developments such as cloning and terminator technology on the agricultural side, and xenotransplantation and stem cell research on the medical side, once these were propelled into the public arena. Once a threshold of awareness was crossed as to the power of modern biotechnology to upset the established biological order, the perceived threat to an established *cultural* (as well as biological) order would almost inevitably create this kind of a backlash effect.

This predisposition was not obvious so long as developments in genetics were presented as "science-as-usual" in popular media accounts. But its eruption was probably inevitable, as genetic advances continued. And it is entirely consistent with the historical fact that the greatest storm of attention, if not actual controversy, that has as yet met any development of biotechnology was that which emerged around cloning technology. The controversy over cloning may have become the "breakthrough" development because it actually threatened an even more deeply held cultural principle, the celebration of the *individual* that is so quintessentially American.

INDIVIDUALISM AND THE GENETIC REVOLUTION

Appreciation of the worth of the individual, the emphasis on individual achievement in a competitive environment, tolerance for individual differences, promotion of individual responsibility and accountability—whether these things take liberal or conservative forms, they are very close to the heart of American culture. This principle has already been established in discussing the importance of individual genetic heritage, but our stress on the responsibility, accountability, and freedom of the individual goes beyond our cultural "take" on the role of genetics. Genes are understood as behaving in a deterministic way but individual human beings still make choices (and take the consequences). And the right to make those choices is very highly regarded in American social and political culture. Commonly this characteristic emphasis is contrasted with social life in Asian countries in which the ties of extended family and community appear very much stronger. The veneration of the individual can be traced in part to the European Renaissance, but in the United States this emphasis seems to have grown particularly quickly over the last century, during which we have retreated from an extended-family to a nuclear-family household and then finally, in many if not the majority of cases, to households consisting of one or two single adults and single-parent families. We emphasize the autonomy of the individual over his or her obligation to the group, and we regularly stress the need for each individual to achieve his or her potential over the necessity of making a group contribution. We see people as having an active obligation to achieve their own advancement.

American individuality, in turn, seems to be tied to our emphasis on individual genetic heritage. Advocates for disabled children argue for their right to achieve their individual potential, a quality for which genes literally provide the blueprint. Little wonder, then, that manipulating human genetic makeup, particularly through cloning, has been seen as a threatening challenge to this concept of a uniquely individual identity tied to genetic structure. While ethicists and psychologists have repeatedly pointed out that genetically identical individuals would not actually be the same person—that is, they would not actually be identical even if raised similarly, and they would certainly not share consciousness—any more than identical twins are the same person, we still seem to be unnerved by this challenge to the proposition that each individual should have his or her own independent roll of the genetic dice. Cloning oneself tends to be seen as a form of personal immortality—an association that confirms the link between identity and genetic make-up. And having a second chance is perceived as being against the rules of nature and of God.

Biotechnology, especially agricultural biotechnology, challenges our valuation of individuality and individual initiative in another way, even though Gans' (1979) study of the values embedded in the U.S. nightly news uncovered the extent to which, as a culture, we continue to value rural life, capitalism, and democracy. While two decades have passed since this original study, during which the American family farm continued to wane in numbers and influence, biotechnology in agriculture is still seen as representing a clash between the economically independent small farmer-small businessman and larger, more anonymous, corporate interests whose public image does not seem particularly benign. This was a large part of the conflict that arose over dairy use of bovine somatotropin. And it is even a larger part of the conflict that has arisen more recently over so-called terminator gene technology that makes it impossible for farmers to save seed from one crop to the next (chapter 8).

In the bovine somatotropin and terminator gene controversies alike, the interests of faceless agribusiness corporations appear pitted against those of our last stalwart bastions of individualism and free enterprise, the small family farmer. Even though farmers who march on Washington to demand favorable treatment in agricultural policy debates seem to gain no more than momentary news media attention, this David and Goliath story of their resistance to large-corporate interests appears to have struck a more sympathetic chord. While some might claim that the impact of agricultural biotechnology on farmers is not radically different from the impact of earlier advances—such as corn seed hybridization, which also required farmers to purchase new seed from corporate sources for each new growing season—it has begun to engender a more critical public response. In part this is probably because of the general sensitivity to genetics-related issues that seems to simmer below the surface of America's romance with science and technology. But it is also consistent with the championing of individual rights vis-a-vis corporate power, and with the more general pattern of risk perception that risks imposed without full and open communication and without choice on the part of those exposed as less acceptable—in fact, as literally more risky.

RISK, TECHNOLOGY, AND DEMOCRACY

What could be more stereotypically an American value than freedom of individual choice? Research on tolerance for technological risks has consistently shown that it matters to us whether risks are openly communicated, whether they are familiar, and whether affected individuals have control over their exposure to a risk (see for example Slovic, Fischoff, and Lichtenstein 1979 and the previous chapter of this book for discussion of related factors). Many segments of the American public are either suspicious of large corporations, distrustful of big government, or both, and they may distrust the motives of corporate interests that are using science for commercial ends. The ecological and health risks of agricultural biotechnology, however small they might be argued actually to be, are risks that are being implemented by these large institutions with relatively little consultation or dialogue with the general public. The kind of coverage found to be characteristic of the mass news media does not tend to raise questions about biotechnology and therefore is unlikely to serve as an adequate forum or stimulus for expanded popular debate. Ironically, only through such debate would the general public likely be enabled to gain the sense of participation and control over outcomes that would be needed for their support.

While limiting public discussion might be seen as serving the short-term interests of biotechnology's promoters, one of the reasons the longer-term consequences are not likely to serve these same interests is that it is public dialogue that creates the opportunity for people to become more familiar and comfortable with risks, as well as benefits, and fosters the perception that democratic processes and popular control are governing collective choices. Of course, such a perception could certainly be cultivated in a manipulative and insincere way as part of a campaign of persuasion, but such a strategy would probably be doomed. The odds are very great that in a country with universal education and some degree of press freedom, the impact of this kind of manipulation would not last. Countries that have experimented with consensus conferences and other mechanisms for gaining public input into the policy-making process for agricultural and other forms of biotechnology actually have a more realistic chance of garnering longer-term public support, difficult as this may be for pro-biotechnology forces in the U.S. to accept. In fact, at least for modern democracies, this is probably the only means available to build reliable public support, even though a particular outcome cannot be predicted with certainty. Even from a strictly self-interested perspective, those who want to promote biotechnology should actively concern themselves with promoting this kind of extended public dialogue. To do otherwise carries its own substantial risk of inducing anti-biotech backlash effects of the type we are already beginning to see.

But the relationship between the opportunity for individual choice and perceptions of technological risk has implications beyond these strategic ones. This relationship is built on a characteristically American valuation of the individual's right to determine his or her destiny. Critics of market capitalism have argued that the consumer's freedom to choose within a capitalistic system has been, at most, freedom within strict limits. Nevertheless, it is a freedom that American consumers hold dear. It should come as no shock to discover that we consider technologies more risky when we have no right of choice to exercise about whether and how they will affect our lives. This is entirely consistent with the "apple pie" cultural values held by most Americans, and is a tenet ignored by biotechnology's champions only at their own risk.

The bottom line does indeed threaten those interests. Advances in genetic technology threaten fundamental features of our worldview, from the very basic distinctions we make between life and death to our reverence for natural biological processes and our valuation of the uniqueness of each individual person. These features of our culture transcend political differences and, albeit in slightly different form, characterize the thinking of environmentalists and traditionally religious people alike. Further, the fact that biotechnology's promoters tend to be large-institution interests can seem threatening to our fundamentally populist political ideal.

Media explanations that present biotech as routinized scientific progress may pacify pro-biotech interests fearful of raising the public profile of related issues and serve temporarily to divert popular attention from troubling questions; they certainly do not serve to amplify fears. But they may also do little to challenge a popular philosophy of genetic determinism that seems to have deeper cultural roots, a philosophy that may prove to be somewhat incompatible with the widespread deployment of genetic engineering and related technologies. And media accounts of this type are unlikely to contribute much to a climate of broad public discussion, the type of climate that is actually a necessary antecedent to public acceptance. (Science fiction may do a better job here than popularized science news.) Even though many of the difficult questions that have been raised about biotechnology are not illusions and should not be regarded as stemming from some sort of cultural distortion, we do live in a culture that is likely to be resistant to this kind of change. Eventual public acceptance cannot be guaranteed, but it is particularly unlikely without extensive public awareness and debate.

NOTES

Some portions of the material in this chapter, particularly the pilot survey and content data, were originally developed in collaboration with Karen Taylor, then a master's degree student in the Department of Speech Communication at Texas A&M University.

1. Issues of free will and original sin aside, however, we hold the individual accountable for his or her actions, once a genetic legacy has been determined; this is one way to make sense of the fact that in our culture capital punishment can be supported even by many of those who oppose abortion.

6

The Labeling Controversy and Public Perceptions of Risk

Because values cannot be evaluated scientifically, they are best addressed by permitting consumer choice at the marketplace through product labeling.

—Marion Nestle, professor and chair, Department of Nutrition and Food Studies, NYU (Nestle 1998)

You've had BST and cookies all your life.

—Advertising copy attributed to biotechnology industry (Roush 1991)

If only because of the way both universities and the professions are organized, people who write and study about the effects of the mass media sometimes seem oblivious to the fact that newspapers, television broadcasts, and magazine articles contain only a fraction of the information delivered to American consumers, especially on nutritional and health matters. An enormous concentration of consumer-targeted messages, both informational and promotional, is to be found on food product packages themselves. These include the now-familiar and official (but once highly controversial) food pyramid from the U.S. Departments of Agriculture and of Health and Human Services suggesting appropriate proportions of foods to be eaten from different categories, to tables listing the nutrient content of each and every prepared product on U.S. grocery store shelves, to carefully worded claims about possible health benefits.

The promotional images that marketers use to wrap each product and each informational message in what they believe is an appropriately enticing emotive context also contain powerful messages, albeit nonverbal ones. This has produced cookies that appear to be baked by forest elves, pancakes that appear to be made by kitchen help straight out of *Gone with the Wind*, and oatmeal that appears to be endorsed by conservative Quaker elders. The context thus evoked is likely to have an important and perhaps definitive influence on how the other, more overt, informational messages are interpreted and understood. It is as much on the product packaging that lines supermarket shelves as it is in the pages of the daily press that the struggle between corporate and consumer interests in the United States takes place.

This book is not about regulation of food labels but about the impact of mass communication on public opinion about agricultural biotechnology. Food labels, however, are a part of the mass communication environment. Themselves mass messages, they are an important source of food-related consumer information in the United States, perhaps the single most important source. And they are also inseparable from the other food marketing messages in which modern supermarket products appear. Food labeling has effects that are not always obvious.

LABELING AND THE RISK AGENDA

Labeling bioengineered products has been controversial, to say the least. The U.S. Food and Drug Administration (FDA) has taken the position that its mandate is to regulate the safety of products, not of production processes. According to this policy, in the absence of evidence that a genetically engineered tomato, for example, is chemically different (aside from largely imperceptible and so far as is known nutritionally inconsequential variations in the DNA structure), there is nothing for them to regulate. Nonfood crop plants such as cotton that may be engineered to repel pests may, ironically, actually receive more federal scrutiny by bodies such as the U.S. Environmental Protection Agency (EPA) because their function is seen as analogous to that of chemical pesticides. Another controversy concerns the enforceability of standards for even voluntary labeling of genetically modified foods (and items such as milk and cheese for which genetically engineered elements may be important in production). Claims that products are free of genetically modified ingredients could be difficult to verify. But it is also feared that such labels would have an agenda-setting impact on consumers, heightening their awareness of biotech-related issues.

Industrial and sometimes governmental interests have argued that since no test reliably distinguishes foods produced with these new biotechnology-based processes from those that are not, labeling is meaningless. This argument assumes in turn that labeling foods as not genetically modified is analogous to making specific nutritional claims or other factual claims that are at least in principle verifiable, rather than the implied and less specific claims associated with the presence of elves, "darkies," or Quakers on product packaging. This is certainly reasonable; the ontological distinction between using genetic engineering and using elfin bakers seems intuitively clear. But arguing against labeling the products of biotechnology on the basis of the challenges in establishing facticity also ignores the fact that the Fair Trade Act already generally prohibits false statements, meaning (presumably) that literal claims about elf baking would be prohibited regardless of whether it was possible to monitor the baking process or test the resultant cookie to determine this. And conversely, more to the point, it ignores the dependence of organic food labels on nongovernmental growers associations to monitor farming practices, not just products, resulting in foods already being marketed as having been grown or produced in particular ways rather than having particular identifiable properties.

Not all aspects of practices considered "organic" result in differences that can necessarily best be established by examining food products once they reach grocery store shelves, nor is their monitoring limited to this kind of post hoc testing. This approach has already been broadly accepted, even though not everyone agrees that organic foods necessarily share any specific, identifiable, product characteristics related to health, nutrition, safety, or even taste. It is little wonder, then, that industry vigorously opposed the exclusion of genetically modified foods from being certified as organic under U.S. Department of Agriculture guidelines. Making an absence of genetic modifications a candidate criterion of organic status not only has the potential of reinforcing public perceptions that these processes are suspicious. It also—and perhaps as importantly—implies that making this kind of a distinction on a product label may not in fact be logistically or logically impossible after all.

While there has been opposition to labeling products as *not* genetically modified on the basis of the fact that such claims cannot easily be tested, there has been no particular opposition to labeling them affirmatively as *being* genetically modified, the strategy first adopted by Calgene in its initial marketing of the Flavr Savr® tomato. Such an affirmative claim is equally difficult to establish. In short, these arguments against labeling (whether voluntary or mandatory) serve the interests of biotech-related agribusiness interests. They may also be self-serving for governmental interests who must manage their work within the constraints of available resources and the existing political and bureaucratic climate. They might be better understood as social products of these constraints and this climate rather than as products of logical deduction or a mandate to protect public interests.

This is not to argue that the FDA's historic position is necessarily either wrong or inconsistent, because none of these analogies is perfect and there is no one simple answer to these questions, but simply to point out that the waters here are fairly murky. This murkiness lurks beneath the surface of the apparent tidy rationality of arguments such as the FDA's "product not process" approach. Neither the FDA, the USDA, nor the EPA has simple or clearly drawn lines of authority with respect to regulating biotechnology; available concepts of these agencies' responsibilities did not anticipate this kind of technological development. Under these somewhat bewildering circumstances, the public's evident hesitation to put their full trust in the ability of current regulatory bodies to manage biotechnology for the public good is rather reasonable; no one seems to claim jurisdiction.

In addition, it is undoubtedly the case that current federal biotechnology policy in the United States—like all policies of all governmental bodies and agencies—is quite responsive to the expressed self-interests of powerful stakeholders. In this case, whether intentionally or not, the absence of a single, specific federal policy or standard concerning the labeling of genetically engineered products and processes has the effect of keeping the public debate narrowly focused on those aspects that most clearly fall under existing regulatory policy. That is, this "non-policy" has a powerful agenda-limiting effect that interacts with the issue-definition impact of mass media accounts that reinforce the status quo by consistently overrepresenting the degree of public consensus and large-institution stakeholder points of view. We have not developed biotechnology policy; we have instead declined to consider it as the novel constellation of issues that it is, and respond to it only when it most clearly fits into a prior agenda. This is an implicitly conservative position and another important, often overlooked, dimension of the way in which U.S. public debate about biotechnology has been reined in.

Because confusion over federal policy has retarded efforts to label foods as non-genetically modified or produced without biotechnology-based processes (such as the use of artificial bovine somatotropin), even though organic products with such labels are finally beginning to appear anyway, U.S. consumers have been robbed of their right to make certain kinds of choices. In effect, the decision as to whether or not to buy genetically engineered foods cannot be made by consumers on the basis of concerns about such things as environmental impact, regulatory adequacy, economic justice, farm policy, ethical considerations (including animal welfare concerns and religious reservations), or a cautious stance with respect to potential health impacts that might be subtle and are not yet proven. Only when and if specific, documented threats to human health associated with specific food products can be identified will the "product not process" approach (for example) dictate FDA action. This is not to assert that any of these other potential consumer concerns are (or are not) justified, only to acknowledge that they exist and that despite the overlapping authority of several agencies of the U.S. government, biotechnology is not regulated in a way that makes it easy to reflect them. Our declining to label genetically modified food products reflects a broader public policy agenda in which technology is not questioned except on health grounds. It is perhaps this difference in policy agendas, rather than in public fears, that most clearly distinguishes the United States from Europe.

Labeling not only acknowledges risks, it serves to define them—to set an agenda of risk-related concerns considered legitimate, and thereby implicitly define others as illegitimate, for purposes of both official consideration and public choice, just as news accounts define some sources and issues as mainstream and others as fringe. Issues surrounding the social distribution of risks, which Beck (1992) has argued constitute a key organizing principle of modern societies, are not a part of our food policy agenda and are not considered reasonable grounds for consumer choice. Question of who gains and who loses among farmers, producers, biotechnology interests, consumers, even government regulators (whose currency is political security), are not defined as being about risks. To the extent consumers can and will ask questions about this and other extended definitions of risk, the more likely it is that they will feel empowered to make choices on grounds that are of concern to them. This is not, in the long run, likely to be good for the biotechnology business.

LABELING AND RISK DISPLACEMENT

Labeling, especially using warning labels, serves not only to acknowledge and define risks but also in a sense displaces them by emphasizing the responsibility of the consumer rather than the producer of a product. Stickers on packages of meat that give details of safe cooking practices are not there solely for altruistic reasons; if the consumer is warned, the producer sees his or her own liability as being diminished. This may be clearest in the case of cigarette labels. We have adopted policies calling for warning labels on tobacco products that present the dire health consequences of smoking, even though we cannot seem to unstick the political logjam that so far has prevented effective prohibition of the products themselves. We shift the responsibility for the consequences to consumers, who are seen as bearing primary responsibility for the assumption of a labeled risk. And in the case of tobacco, indeed they have, despite a recent series of lawsuits against tobacco interests by individual states that recognized that the costs of smoking are actually collectively shared in the form of much higher medical and insurance costs.

The same dynamic of risk displacement characterizes nutritional labels on food packages. These are without doubt a potentially valuable form of consumer education, and they certainly serve to enhance awareness of nutritional issues, even though existing regulations let sodium be described as a nutrient and do little to explain the importance of the presence of certain kinds of fats. Yet the end result is that most responsibility for nutritional adequacy and balance and the avoidance of harmful ingredients is displaced onto the consumer; producers are more or less off the hook. Market forces—that is to say, consumer choices—have at least resulted in lower fat items appearing in the stores, even where the reductions have not been adequate to make the final products particularly healthful. But in the process, nutritional labels have become as much a marketing tool as an educational one. Foods that can be legally labeled as lower in fat or free of cholesterol are not necessarily foods that a nutritionist would identify as being healthy choices, and it is difficult to know with certainty whether the typical consumer has been helped more than harmed in the process. Indeed, some popular diet books now proclaim that fat is not even the problem and that it has actually been oversold as a health risk; whether accurate or not, the marketability of such material tends to bolster the assertion that food consumers have by now become suspicious of mainstream nutritional claims.

Despite these complexities, labels do acknowledge the existence of risks, sometimes reflecting and even reinforcing a broadening of the public risk agenda—just as news media can reflect but also reinforce public opinion in other matters. Smoking and the excessive consumption of fat are acknowledged as leading contributors to early deaths in the U.S., and we have responded to this partly by labeling, as though letting at least knowledgeable consumers make informed choices solved the problem. But we still expect the individual consumer to be responsible for the final decision, as well as its consequences. The language of freedom that surrounded the tobacco debate in the U.S. Congress in recent years is instructive. Our valuation of individual choice translates into such powerful rhetoric that we seem to have accepted the argument that outlawing an addictive substance that gives people fatal cancers and greatly increases deaths by heart attack is an infringement of the rights of individual citizens.[1] Yet we resist giving citizens the right to choose between genetically modified and unmodified foods because it would be inconvenient and expensive.

If labeling can be used to help displace a risk from a food (or tobacco) producer onto the consumer, why is it widely opposed for biotechnology? Labeling represents a political compromise when we are unwilling or unable to eliminate a *generally acknowledged* risk, and then it does have the effect of moving the burden of that risk (and to some extent perceptions, at least, of associated liability) from the producer to the consumer. But agricultural biotechnology interests have opposed labeling because it seems to represent a tacit acknowledgment that some risks may in actuality exist, and once in place it also allows consumers to make choices that might be based on a wide variety of grounds, not just a narrowly defined set of specific health and safety criteria policed by our current bureaucracy. To adopt labeling of genetically modified and other biotechnology-based food products is both an acknowledgment that some things about these products might be risky and an invitation to consumers to in fact "vote with their pocketbooks" about the acceptability of those risks, an option presently unavailable. Labeling is used to deflect criticism from vested tobacco interests; not labeling is used to deflect attention from the biotechnology debate and thus protect a different set of interests. Not only is this bad democracy, but it leaves public suspicions simmering underground rather than addressing them in public, robbing consumers of their sense of control and in all likelihood exacerbating their fears in the long run.

LABELS AND PUBLIC RELATIONS

Despite the existence of these kinds of powerful arguments for and against labeling of genetically modified foods and other products of agricultural biotechnology, and regardless of the number of opinion surveys that have been conducted on these and related topics, relatively little empirical evidence bears directly on the impact of food product labeling. We can infer, from the media agenda-setting research, that the presence of labels identifying foods as genetically modified will in fact direct people's attention to the issues of risk that might be associated with biotechnology. As has already been discussed, the food industry's fears here are understandable. But as the public debate about biotechnology widens and deepens, consumer awareness of allegations of risk is increasingly inevitable. And it does not necessarily follow that directing attention to risks always translates into heightened sensitivity to their magnitude, although this is commonly assumed even by some of those doing research in this area. Rather, open discussion of risks can also enhance people's senses of familiarity, choice, and control, which is more likely to reduce than to amplify perceived risk.

Even from a biotechnology industry perspective, little may be lost and potentially much may be gained by acknowledging that questions in this area have more than one legitimate side. We can reason on the basis of the persuasion literature over the last fifty years that labeling can permit more open consumer choice without necessarily harming biotechnology-related interests, although this line of reasoning also invites us to acknowledge the potential for labeling to serve a variety of hidden industry agendas. But in the end, the decision is at least as much a matter of ethical, political, and market theory as it is an empirical one.

Having said that, there is some empirical evidence that is relevant to predicting how consumers are likely to respond to information about foods produced using the new genetics. Some of this has already been discussed in the context of public opinion formation. Lay assessments of risk follow a calculus that is more complex than that of the risk assessor; threats, hazards, and uncertainties are unlikely to be cleanly separated, small probabilities of major catastrophe weigh more heavily than their likelihood alone would dictate; and factors such as faith in regulatory mechanisms and perspectives on ethical questions are inseparable from the evaluation of risks. As Thompson (1997) argues, the categories of impact that we treat as analytically distinct from both a regulatory and a scientific perspective (social, environmental, and health impacts, for example) may not be perceptually distinct for nonspecialists, implying that a specialist's bias toward treating them as distinct may not serve nonspecialists effectively. If the result is consumer alienation from the food and further erosion of trust, perceptions of risk will rise concomitantly.

Additionally, Frewer, Hedderley, Howard, and Shepherd (1997) argue on the basis of data from 400 U.K. respondents that not only should effective communication about genetic engineering be application-specific, it should also address questions of environmental impact. While it is unknown to what extent their results might translate to U.S. respondents, the clear distinction that emerged in this study between reactions to human and animal biotechnology versus plant or microorganism biotechnology is consistent with the results of U.S.-based survey research discussed earlier, suggesting that the thinking of these two populations may not be far off. If Frewer, Hedderley, Howard, and Shepherd's results are correct and applicable to other populations, consumers are going to want information about more than just risks in the narrowest sense of human health, whether delivered on product labels or in the mass media. If environmental concerns are not so acute in the United States as in the United Kingdom, they are unlikely to be very far behind them.

We know of few reliable ways to alter strongly held values, attitudes, and beliefs. Whatever the American public concludes about agricultural biotechnology, those conclusions will be rooted in those elements of our collective psyche that are likely to be quite resistant to change. However, even those who believe that opinion in this area is nevertheless malleable still argue that open communication is likely to have the most positive possible effect on public perceptions, even where it might sometimes lead to polarization of negative as well as positive opinions (Frewer, Howard, and Aaron 1998). Boulter (1995) argues that scientists have a responsibility to engage in public debate on these issues; Hillers and Löwik (1998) argue that Dutch resistance to the introduction of genetically novel foods probably resulted more from the sense that such products were being "pushed" onto consumers without their knowledge or consent than from other reasons. This case has a vital lesson for the United States. Once fully aware of agricultural biotechnology and the associated controversies, it seems unlikely that U.S. consumers will respond any differently. No amount of "information" on the safety of bioengineering will then be likely to quiet fears that are rooted in the perception consumers have been deceived rather than in responses to product riskiness per se.

Grunig (1989) offers an analytic distinction between public relations strategies that is useful for analyzing the case of food and other agricultural biotechnology. He identifies four basic perspectives on what public relations is about: "press agentry," or propagandistic attempts to garner media attention by any means available; the public information approach, or the dissemination of information deemed accurate but with the accent on sympathetic results; "two-way asymmetrical" approaches, involving the gathering of public opinion information in order to better manipulate the public, sometimes justified as being "for their own good"; and the "two-way symmetrical" approach, in which genuine accommodation is sought between a public relations client and the public that client is trying to reach.

Grunig argues that the fourth or "symmetrical" approach is not only the most socially responsible, it is the most effective. Labeling products of the new genetics in the marketplace, even if responsive to the broad range of concerns that seem to exist in consumers' minds, could be done in either an asymmetrical or a symmetrical way. That is, real responsiveness to public concerns should be distinguished from the manipulative use of product information to produce a positive reaction. This kind of manipulation, Grunig argues, is not likely to be effective. For the American public to accept biotechnology, labeling and marketing strategies that reflect a real responsiveness to their misgivings will be a better tactic, even if this means some specific applications will be foregone. And these decisions are more likely to develop from corporate rather than from governmental initiative because of the current diffusion of relevant governmental responsibility across several agencies.

The shape that such a response will (or might best) take remains to be determined. Product labels will likely have a role, but it is not necessarily practical or useful to try to use them to respond to the full range of issues that have been raised. So labels indicating biotechnology-based content will continue to depend on broader information in the press and other media contributing to an atmosphere of informed public debate about whether and how this content matters. Here the interests of the industrial developers of biotechnology and the interests of democracy actually converge; long-term public support for biotechnology is only likely to be generated in an atmosphere of open and symmetrical communication. Relying on public complacency rather than seeking to respond to public concerns is probably the most risky strategy of all, as public responses to both animal cloning and terminator genes have recently demonstrated.

LABELING AND DEMOCRATIC THEORY

Social theorist Jacques Ellul has noted (1965) that the most effective propaganda is true, or at least it is composed of information that cannot be proven to be false. Food producers, like national leaders, do not need to make blatantly false statements to persuade their audiences of their messages; it is more convincing to use accurate ones. However, decades of communication research have also consistently shown that two-sided messages are more persuasive than one-sided messages for almost all audiences. Unless an audience is unsophisticated and unlikely to ever hear the other side of the case, an increasing rarity in the modern world, messages that acknowledge the opposing point of view and then provide reasonable grounds for rejecting it are more powerful.

From this perspective, there is no particular reason for biotech's promoters to fear public debate. As in any public relations context, it is better to acknowledge negative information and opposing arguments than to be perceived as suppressing them. It would certainly be better democracy. Whether or not they might have been effective a few short years ago, one-sided messages about biotech's fruitful promise—*however accurate*—are unlikely these days to be persuasive. And while product labeling might be feared because it can direct the public's attention to issues that might otherwise have been widely ignored and because it has the power to define these issues as representing risks, openly identifying products as containing genetically modified ingredients could equally well help to dampen rather than exacerbate negative public opinion.

Labeling products with risk-related information may actually tend to convey to consumers that the risks are much more fully known, predictable, and measurable than they actually might be, and in addition that they are safely under official regulatory control (whether they actually are or not). Labels certainly also convey that risks are being openly communicated in a straightforward way, which is known to be an important component of public reactions to risks. In the long run, the availability of this information will very probably have a desensitizing effect on people's general concerns about bioengineering. In other words, under many circumstances adding labels, even labels that warn of risks, may serve the interests of biotechnology's producers and marketers as much as consumers. For a variety of reasons, the biotechnology industry might not need to fear them as much as it does.

However, while the potential certainly exists for food labels to be used in manipulative ways, product labeling that provides consumers with information that they want most clearly benefits consumers. It empowers them to make informed and reasonable choices, on whatever grounds, and it is therefore in the interests of democracy. At the same time, if there really is nothing to fear associated with bioengineered foods, then open communication is a win-win situation for consumers and industry—consumers gain power to make wiser purchases, and producers who have a better choice to offer gain market share. Only if there really is something to hide associated with biotechnology does the industry stand to lose in the long run. Conversely, this is precisely the reason that appearing to have something to hide is such an extraordinarily poor public relations strategy. It is entirely possible to argue for biotechnology product labeling on the grounds of enhancing free choice, free markets, and open democratic process alone. It is also possible to make this argument on behalf of biotech's commercial stakeholders, who probably have more to lose than to gain by continued opposition. However, at present neither point of view can be fully supported by social science. The former is more a matter of philosophy than an empirical question, and the research needed to adequately explore the actual impacts of labels on consumer attitudes in the case of biotechnology-based foods largely remains to be done.

NOTES

1. I am not arguing this was necessarily wrong, merely that it incorporates cultural values. The extent to which government should interfere in individual decision-making is, of course, a complicated question—and one outside the scope of this discussion.

7

The Cloning Story

In theory, . . . such techniques could be used to take a cell from an adult human and . . . create . . . a time-delayed twin.

—Gina Kolata, *New York Times* science reporter,
front-page story on February 23, 1997

The discovery by a previously obscure Scottish animal scientist named Ian Wilmut of a technique for cloning lambs using DNA from the cells of an adult sheep dominated U.S. media coverage during early 1997 in ways few other news stories, and perhaps no other science story, had ever done. Furthermore, the story was about disagreements over ethics, a topic category usually treated gingerly if at all by U.S. journalists. Both the norms of professional journalistic practice and a particularly American cultural perspective contributed to the prominence of this story, creating a novel news frame in which the opinions of ethicists acted as counterpoint to those of scientists. But broader institutional interests also helped shape this debate, which in the end served primarily to maintain the existing distribution of power rather than subvert it.

This chapter reviews key characteristics of elite U.S. newspaper coverage leading up to, during, and immediately after the flurry of publicity surrounding Wilmut's discovery, with a particular focus on stories discussing the ethical dimensions of the work. This case study of cloning news illustrates how both these factors (professional practice and the cultural system in which it is embedded) shaped the resultant controversy. Nevertheless, these two factors together do not produce a complete account. The particular types of ethical arguments put forward also represent a rich source of perspective on how other institutional interests may have influenced the debate.

Although the mass news media (along with science fiction) are among the few forums that are available in the United States for something resembling open public debate, scientists generally dislike their work appearing on the public stage. They fear its being misunderstood and are anxious lest misrepresentations of fact should exacerbate tensions. Various commercial interests with stakes in public opinions and perceptions about science and technology may also fear publicity and what they see as the exaggeration of risks. But without this debate, neither the interests of society nor the interests of science are likely to be fully served; science policy will remain something made behind closed doors in ways that are invisible to—and therefore not trusted by—voters and taxpayers. In this sense, media attention to controversies over science-related issues such as cloning, mad cow disease, use of the pesticide alar on apples, and so on is generally more constructive, from the points of view of both science and business, than might be assumed, this point being made in the previous chapter concerning product labeling. This is, in part, a case study in the role of mass media as a stimulus for broader public discussion and in the limitations of commercial media in that role.

The public obsession with the ethics of cloning is also a remarkable resource for understanding the way in which our culture privileges some ethical issues over others. Because the debate was inspired by a development in science (or, at least, technology), its vigor is doubly unusual. Under other circumstances, U.S. science journalists generally seem committed to divorcing scientific facts from their social, political, and economic context. Ethics are rarely news; there seems no "objective" way to report on them. Nevertheless, it is remarkable how the news media's cloning debate was largely constrained to the *biological* dimensions and their implications for *human* reproduction, highlighting some ethical issues but ignoring those surrounding the economic implications of biotechnology, including the likely impacts on agriculture, as well as implications for the integrity of both nonhuman animal and plant species and for ecological health and balance. These other issues were not unknown at the time; they in fact arose sporadically only to immediately submerge again in discussion of the ethics of cloning human beings. They were not discounted so much as momentarily drowned.

While media content is not a direct indicator of public opinion, it does reflect—and therefore can reveal—deep-seated cultural values, including our belief that capitalism is inherently benign (Gans 1980). Almost by definition, this is a belief that supports the status quo distribution of both power and benefit against claims for alternative forms of distributive justice and "common good" claims for environmental protection. Rejecting the stronger claim that news accounts might be part of a conspiratorial *intent* to preserve status quo privileges does not mean that these accounts are without significant effect vis-a-vis the distribution of power. Mainstream media research has amply demonstrated the power of agenda-setting effects (McCombs and Shaw 1972; Iyengar and Kinder 1987; McCombs 1992), and public opinion research suggests a way that media accounts can create a spiral of silence without necessarily changing people's actual opinions (Noelle-Neumann 1993). Albeit perhaps unwittingly, then, news can indeed serve a particular set of political and economic interests—those of the status quo—by reinforcing certain "definitions of the situation" consistent with ambient cultural beliefs while ignoring others.

The remainder of this chapter is based on a review of relevant elite U.S. newspaper coverage indexed in the National Newspaper Index[1] for the years 1994 through 1997 conducted for the purpose of exploring these phenomena, using news items and newspaper commentary as a resource for understanding how this single issue (human cloning) came to dominate public discussion, serving as a "crystallizing symbol" (Edelman 1971) for public concerns about biotechnology while at the same time rendering nearly invisible—and thus delegitimizing—alternative viewpoints raising different objections. Keyword searches used the terms "biotechnology," "food biotechnology," "agricultural biotechnology," "animal cell technology," "plant biotechnology," "ethics and biotechnology," "livestock," "genetic engineering," and "cloning"; the latter two terms identified the great majority of articles considered for inclusion in this study. From this initial screening, 130 articles whose titles suggested at least some concern with ethical implications were selected for further analysis. No particular claims are made that this research is based on an entirely comprehensive subset of U.S. media coverage—even elite media coverage—of biotechnology during this period; for a more systematic and quantitative treatment not especially focused on representations of ethics, see Lewenstein, Allaman, and Parthasarathy (1998).

CLONING AND SCIENCE JOURNALISM

Some of the earliest accounts of Wilmut's accomplishment were glowing. Thomas Maugh of the *Los Angeles Times* called it "an unprecedented feat," a "remarkable feat," and a "feat [that] should also have marvelous applications," all in a single story (1997, A1, A16). But journalism thrives on controversy despite a strong tradition of "gee whiz" science reporting, and it was almost inevitable that a science story that didn't die would also be a science story not centered on the accomplishment per se. The exact nature of the controversy as presented in U.S. journalistic accounts was very much an outcome of the way journalism in the United States is practiced, involving procedures in which the interaction of journalists and their sources creates a particular frame or definition that quickly becomes the lens through which any given story is inevitably filtered.

In recent decades much effort in media research has shifted from studies of the "effects"[2] of the media to a focus on the organizational routines (Tuchman 1978) and the multiple other cultural, structural, and ideological influences (Shoemaker and Reese 1996) that produce media content. A key characteristic of the practice of science journalism is the dominance of an objectivity ethic that draws from both scientific (Nelkin 1995) and journalistic (Schudson 1978) professional traditions. In the United States, ironically, prominent science news stories often become sensational through either overinterpretation or misapplication of this "objectivity" ethic.[3] These processes determine how material from sources, especially expert sources, is presented.

Sometimes this takes the form of treating speculations as fact. For example, the 1990 earthquake "prediction" for New Madrid, Missouri—produced by an economic forecaster who had taken up the practice of modeling weather patterns—was widely reported as a legitimate scientific conclusion, engendering considerable media attention and substantial public concern. The 1989 announcement by two Utah scientists that they had produced a cold fusion reaction in their university laboratory was perhaps more reasonably treated as scientific but still fell short of being fact, a circumstance media accounts were widely believed to have ignored (Lewenstein 1995). Reporting assertions cloaked in the trappings of science as prima facie scientific masks controversy and uncertainty.

It is likely this same kind of uncertainty underlies public frustration with mass mediated nutrition and health information. When the results of each individual study (as disseminated by the public information activities of universities, academic journals, and other reputational stakeholders) are reported as factual scientific conclusions in and of themselves, science appears to be reversing its position constantly on all fronts.

At other times, however, objectivity for science news, as for news of other kinds, is defined as the production of a balanced account. Presumably this interpretation of the objectivity ethic derives from political reporting, where a "left" and a "right" position are always (simplistically enough) assumed. But this practice can also be destructive. In the case of the global warming controversy, for example, it seems to have led journalists to seek out (in order to balance their stories) maverick opinions that discounted the mainstream scientific consensus. Whether deliberate or inadvertent, this practice tends to legitimize those maverick perspectives. In the case of global warming, acid rain, and other environmental issues, the result is sometimes media accounts that offer "scientific" justification for political paralysis: consensus is never achieved.

Concern for fairness and balance also led journalists covering the recent reemergence of the creation "science" debate in Arkansas to present creationism and evolution as competing theoretical interpretations of available evidence without regard for creationism's lack of status as a *scientific* theory (Taylor and Condit 1988). Such issues can present enormously complex challenges of fairness, interpretation, and presentation not easily resolved on a short deadline. Which perspectives must be discounted? Can journalists legitimize all points of view without misrepresenting science in general? What about situations where the science itself is uncertain?

Of course, maverick scientific opinions can turn out to be correct (that is, can later become the reflections of reconstituted scientific consensus). This is not confined to the era of Copernicus. Early in the history of the emergence of the HIV theory of AIDS, alternative medical explanations of the disease were more common. Post hoc, these alternatives do not seem to have garnered much evidence or support within the medical research community. But at the time, before it was clear which lines of research would eventually come to be considered most productive, credentialed scientists pursuing alternative theories had difficulty getting news media to take them seriously, and concomitant difficulty getting funding. The information subsidy activities (Gandy 1982) of established institutional interests are enormously powerful determinants of how scientific facticity is defined. Would we be closer to, or farther from, preventing or curing AIDS if these processes operated differently?

Confusing the public is not the only issue. Not only are scientific reputations and the reputations of academic and other research institutions at stake but so are millions of dollars of research funding. When commercial interests are added into the equation, as is the case for biotechnology, the stakes get vastly higher. Generally speaking, print media coverage of biotechnology is heavily dominated by industrial-commercial interests (Priest and Talbert 1994; Lewenstein, Allaman, and Parthasarathy 1998). And for better or worse, popular media accounts of science do in turn influence academic research agendas, especially in interdisciplinary areas (Clemens 1986). Journalists' reliance on institutional information subsidies is a key characteristic of professional practice that systematically "colors" the news according to the interests of those sources.

This phenomenon is not limited to the activities of commercial entities. If anything, the cloning story is unusual because it is dominated by academic researchers and ethicists, along with a handful of theological voices, rather than strictly commercial interests. Information subsidies are also provided by interest groups and other nongovernmental, nonprofit organizations, and the self-interested activities of media organizations also have influence despite the objectivity ethic. The use of alar on apples is widely recognized as having been brought to public attention by consumer activists and publicized by the television news magazine *60 Minutes* in ways that propped up the broadcast's image as a forum for original, hard-hitting, exclusive, investigative stories. We can only hope that social responsibility and public interest objectives can coexist with—and be well served by—the organizational self-interests of news media.

Both Wilmut's Roslin Institute employers and the scientific journal *Nature* (in which Wilmut's work was literally just about to be published when the mass media story broke) had stakes in making Wilmut's sheep-cloning accomplishment as prominent as possible, and both acted accordingly by making material and sources available to the press in the first place. Their collective actions placed the story on the news agenda to begin with. While not so readily recognized as commercial stakeholders as are large agribusiness corporations, both research organizations (within and outside of the academy) and scientific publishers are also stakeholders for this issue. Individual journalists such as Gina Kolata of the *New York Times* also had career stakes in seeing their work widely read and recognized.[4]

Finally, activist groups, like political parties, stage "pseudo-events" designed as much to garner media attention as to accomplish any other organizational purpose. This tactic may be the only way fringe groups (representing opinion considered outside the current mainstream) can get coverage at all; the early history of Greenpeace's environmental activism is probably the best-known example. While the coverage may have tended to delegitimize Greenpeace by branding it as extremist, the organization's tactics were nevertheless effective in drawing public attention to their concerns regarding whaling activities and later other environmental issues. Similarly, anti-biotechnology groups, such as Jeremy Rifkin's Foundation for Economic Trends, have had a voice in media coverage of related issues, although—as for Greenpeace in its early years—they also tend to be pictured as extremist. Rifkin's tactics are not especially extreme, but his group's position—general opposition to all genetic engineering—is perceived to be.

Fringe groups' voices can be used to balance stories, but they tend to be tacked on at the end as though they were afterthoughts, a practice that contributes to the impression they are to be taken less seriously than other interests. Such sources also tend to be described in terms of their role in the controversy ("anti-biotechnology activist") rather than their credentials, another delegitimation tactic. However, perhaps because of the ready availability of academic bioethicist sources, Rifkin's voice was not particularly prominent in mediated discussions of cloning. Since many of the academic bioethicists quoted are associated with medical programs, it is likely they helped keep the discussion centered on issues of human reproduction.

CLONING AS A CULTURAL PHENOMENON

Coverage of cloning involved a number of features characteristic of science stories in U.S. media, in particular the interpretation of objectivity as balance between opposing points of view and the dominance of commercial and other interested sources. Stakeholder institutions played influential roles and opposition voices were simultaneously used to meet the balance requirement and yet frequently delegitimized. But cloning was unusual not only in the prominence of the story, but also because—unlike many of the examples above—the controversy was not over the scientific facts (even though individual stories varied in their representations of these facts), their interpretation, or even their implications for policy per se. It was over the ethics of making particular use of those facts.

This might have been partly because Wilmut's discovery itself was primarily technological (having to do with the technique or process for cloning new individuals from adult cells) rather than scientific (adding to our stock of theoretical knowledge). This was not immediately recognized by the science journalists covering the story, to whom Wilmut had to explain with considerable patience and modesty why it was unlikely he would win a Nobel Prize for his work. Nevertheless, the journalists persisted in covering this work as a science story. Controversy did emerge, but it was over the appropriate ethical response rather than over scientific facts—specifically, whether the technology should or should not be employed and in what circumstances.

The emergence of public controversy over the application of a technology is almost unprecedented in the United States. Some, although not all, of the appeal of this frame for the story can be explained by American's cultural obsession with genetic heritage and individual genetic identity (Nelkin and Lindee 1995). Cloning, especially the prospect of cloning human beings, presented a profound challenge to our deeply cherished notion of individual human identity. We are also, however, a culture with a profound belief in the equation of new technology with human progress. We are resistant to legislating technologies at all, although this has been tempered in recent years by a growing consciousness of environmental impacts. So cloning presented a bona fide dilemma for the United States from a cultural perspective.

The opposition of the Rural Advancement Foundation International to Roslin's patenting of Wilmut's cloning process received some *Wall Street Journal* coverage, but no explanation; an Associated Press story mentioned only that concerns about cloning included "fears about the potential for cloning humans" (Scottish Institute 1997, B2). Other arguments appeared briefly only to be discounted or simply dropped. Perhaps reflecting its business orientation, the *Journal* had run a piece on February 4 that called India's government-mandated destruction of an unlicensed test plot of engineered eggplant the work of "scaremongers" and "divorced from any scientific basis" (Bitter Fruit 1997, A18). That cloning technology broke this pattern of dismissal is surprising, but the pattern of attributing opposition to "fears" did not change.

The emphasis on the immediate prospects for cloning human beings (as opposed to animals or even plants) was a media phenomenon best understood as a result of cultural factors rather than professional practice per se. This definition of the issue, or frame, was applied to the story immediately and consistently, despite the fact that it had relatively little to do with Wilmut's or the Roslin Institute's plans. In a typical account, *Wall Street Journal* staff reporter Robert Langreth wrote on February 24, "Theoretically, there isn't any reason why the new method wouldn't work in other mammals—including humans. . . ." (1997, B1). This type of coverage occurred despite Wilmut's consistent denial that any such effort had been contemplated or would even be considered acceptable. The persistent denials are likely a reflection of the frequency with which this question was asked.

Perhaps ironically, the concern over this specific issue (the deliberate cloning of individual adult human beings) may have originated partly from the ethicists themselves. Langreth goes on to attribute to University of Pennsylvania bioethicist Arthur Caplan the point that no U.S. laws existed to prevent cloning people. Caplan is also quoted in a February 24 *Los Angeles Times* story in which he asserted that even to attempt to clone a person is "outrageously immoral" (Hotz 1997, A14). While balanced in this case by comments from John Fletcher of the University of Virginia to the effect that under some circumstances (such as in attempts to avoid passing on an inheritable disease) cloning might not ethically be out of the question, the general tone of this and other similar articles is that the ethicist community is both outraged by the prospect and convinced that human cloning is a likely future scenario.

As framed in journalistic accounts, then, ethicists' perspectives served to exacerbate fears about carbon copy human reproduction rather than to calm them or to raise other, perhaps less sensational, issues. In the process, a few individuals—including Caplan—became momentarily prominent, assuming a role analogous to the one that has been called that of the "visible scientist" when it occurs within the scientific community (Goodell 1977). Wilmut himself, momentarily elevated to international stature, when asked repeatedly about the prospects of cloning people, and apparently sincere in his opposition to any such effort, may also have contributed to this emphasis by his very defensiveness on the point.

In news coverage of cloning, however, ethicists' comments—like those of biotech's opponents, when the latter appeared at all—were also frequently used as balancing material in stories otherwise presenting a monolithic scientific point of view. This was true despite the fact that the ethical comments were often deeply buried toward the end of the story. Even so, the presentation of ethical arguments in this kind of juxtaposition to scientific ones both satisfied the journalists' objectivity criterion and created stories with some controversy even though the facts were not themselves controversial. Instead, scientific and ethical arguments were pitted against one another. Sometimes the ethical arguments were delegitimized by placement or emphasis. One of the early *New York Times* accounts, on February 23, was a story dominated by comments from Wilmut and other animal scientists but rounded out in its closing paragraphs with comments by Robert Munson, University of Missouri medical ethicist, who talked about cloning Jesus. Munson's intent was presumably to illustrate issues of personal identity, but this needed to be read into the story. Including the arguments of ethicists, especially those likely to strike a reader as extreme, allowed journalists to produce "formula" controversy stories incorporating divergent expert points of view even though there was little controversy available within the scientific community. Perhaps incidentally, the emphasis on the more polemical ethical arguments invited dismissal. It certainly portrayed ethics as pertaining to relatively bizarre scenarios.

In our culture, ethical reasoning is not seen as having the force of scientific reasoning; ethical conclusions are classified as judgement or opinion, rather than as factual or explanatory, and it is typically access to facts—specialized knowledge—that justifies claims to expertise in our society. So it is not surprising that media frames sometimes tended to treat ethicists' comments as coequal with those of scientists (that is, as expert opinion) and sometimes did not. Social scientists' conclusions are also presented as opinion (Evans 1995). Nevertheless, it is unusual to find ethical debate so prominent in U.S. media accounts in any form. Coverage of behavior adjudged unethical is common; coverage of debates about what should and should not be considered ethical is much less easily framed as objective reporting and therefore is much more rare. Cultural factors and journalistic tradition converged to produce this outcome in the case of cloning. But these explanations are not entirely complete.

CLONING AND INSTITUTIONAL INTEREST

In addition to the active participation as actors, sources, and "spokespeople" by immediate stakeholders in the production and support of the news agenda (researcher Wilmut, Kolata and other key journalists, Roslin Institute, and *Nature*), the agenda-building process draws in and reflects the influences of other institutional interests. Political leaders, for example, seek to establish their competence by demonstrating responsiveness to emerging issues: President Clinton did so by asking the already constituted National Bioethics Commission, which had up until this time received little public attention for its activities, to consider the question of cloning, making his administration appear to be immediately on top of a visibly emergent situation and delivering to the news media an entire panel of potential expert sources on ethical issues. Even ethicists were hardly immune to seizing an opportunity to have themselves and "their" issues thrust into prominence; a handful were quoted repeatedly.

The scientific community, most visibly in the persona of the director of the National Institutes of Health, reacted slowly but predictably to the perceived threat of federal regulation by urging Congress to be cautious about passing legislation restricting research activities, underscoring the potential medical benefits of further development of cloning technology. Clinton's demonstration of administrative leadership had had the effect of underscoring the perceived ethical significance of Wilmut's discovery, necessitating substantial damage control from scientific leaders within his own administration.

The biotechnology industry, however, which might have been expected to be the interest group most directly concerned with the outcome of this debate, was strangely quiet, making no particularly visible attempt to "spin" the issue in a favorable direction. Instead, everyone seemed united in opposing human cloning. In order to fully understand why this would be so, it is vital to recognize that defining the cloning issue as a question of the ethics of human reproduction itself served the interests of the biotech industry. Human reproductive cloning has not yet spawned an industry and is not immediately likely to—the intentions of a well-publicized maverick Chicago physicist named Richard Seed, who planned to be the first to clone a human being even if this meant moving his research outside U.S. borders, notwithstanding. So the emerging public outcry against it did not threaten established interests in any important way.

Unlike other forms of biotechnology involving plants and animals, cloning technology was not perceived as a particular environmental risk or an economic threat to agriculture. Human cloning is actually a fairly "safe" issue from the point of view of the burgeoning biotech industry.

While concerns in Europe and elsewhere over other forms and aspects of biotech were covered during this same period, they were sometimes represented as fringe positions or, in at least one case, a matter of cultural eccentricity. A 1996 Associated Press piece carried by the *Wall Street Journal* had described the Germans as being "furious" that the European Parliament would not ban genetically engineered ingredients that might impair the purity of their beer (Germany gets 1996, A5C). A *New York Times* editorial conceded that altering the human gene flow might be dangerous, but borrowed industry's refrain to the effect that cloning would be "only a step" past current, widely accepted, selective breeding practices (Cloning for good 1996, A26). Meanwhile, biotech's presumed economic potential and probable contribution to resolving world hunger problems were rarely challenged, and never persistently so.

Dimensions of animal cloning involving animal welfare and animal rights issues were also poorly highlighted at best in this debate. While the prospect of cloning people might entail some of the same concerns about playing God and the commodification of life that have been raised for other forms of biotechnology, the cultural distinction between human and all other forms of life is so clearly drawn in our society that arguments about cloning people do not easily spill over into other realms. Related arguments were raised early on; for example, an unsigned *Los Angeles Times* piece on February 25 pointed out that "genetically identical" livestock would be more susceptible to disease (Next 1997, B6). But these did not persist as themes.

The cloning debate did not necessarily come to focus on the ethics of human reproduction *because* this set of arguments did not directly threaten the established biotechnology industry. Professional journalistic routines and practices acting within a particular cultural context were what framed the debate in this way. The prospect of cloning individual human beings had profound resonance with our cultural emphasis on individuality and individual genetic identity. Framing this theme as a conflict between science and ethics reflected practices long characteristic of science journalism, as can be readily illustrated for other recent issues. And other institutional stakeholders shaped the debate. However, at the same time, it is difficult not to speculate that the outcome would have been different had other, more established, institutional interests been seen as more closely tied to the issue. Only the interests of the scientific research community, concerned about a potential Congressional overreaction that might put restraints on their work, seemed to be at stake; this stake was not visible immediately, and scientists have historically been weak defenders of their own interests. These arguments were, for the most part, not raised until after the cloning frenzy had already begun to wane. Perhaps the research community was, as usual, simply not anxious to stir up additional public attention.

The existence of institutional interests generally tends to exert a dampening effect on journalism, and while the precise degree of influence of such factors is difficult or impossible to establish, mainstream editors and publishers are not anxious to offend powerful people and organizations. The "watchdog" tradition notwithstanding, battles must be chosen carefully for news organizations to survive and thrive. Concentrating the public debate on the human reproductive applications of cloning might appear at first glance as an exception to this rule, but in fact it implicitly served the interests of the biotechnology industry by confining the debate to this one dimension. It was a "safe" controversy. Understanding the journey from reports of a discovery of a technology for cloning sheep for the purposes of pharmaceutical production in their milk to a public debate focused on the ethics of cloning human beings requires sensitivity to all three factors: professional practice, cultural context, and institutional interest. Like water moving downhill, and quite contrary to our conventional beliefs about the political role of journalism, media attention flows along lines of least resistance. This was as true for the coverage of the Nixon Watergate affair (Lang and Lang 1983) as it is for today's coverage of biotechnology: journalists follow controversy much more often than they create it, conventional wisdom notwithstanding. The ideology that says otherwise, that relies on journalism to ferret out issues not otherwise visible to the public eye, implicitly supports established interests, if only by nurturing a false sense of complacency.

Nevertheless, a debate of sorts did take place, and public attention was focused on a few of the issues surrounding contemporary biotechnological developments, albeit an abbreviated set of them. While journalistic practices, cultural predilections, and structural dynamics constrained the presentation in substantial ways, ethical issues were highlighted and some public discussion that might not otherwise have taken place ensued. The debate, however, was not one likely to challenge existing institutional arrangements or issues; however compelling to some, the cloning controversy was essentially harmless to the status quo. Throughout this period, objections to genetic engineering for agricultural or pharmaceutical (rather than human reproductive) purposes on economic, environmental, or other grounds were nearly invisible. This did little to prepare the United States for the ensuing trade war with Europe over genetically engineered foods.

Instead, journalistic practices such as reliance on available expert sources created the opportunity for institutional interests to frame the issue in ways that played upon ambient cultural predilections. The effects of this kind of framing or issue definition may be limited (Priest 1995), but they are not without consequences for setting the public debate agenda. Understanding media accounts of the cloning controversy provides us with a window that allows us to observe the interaction of culture and structure, as well as the way mainstream journalism "props up" status quo power relationships rather than challenging them in significant ways. Here culture becomes an ideological resource for maintaining structure, in part through the ways in which journalism is practiced.

NOTES

1. This chapter is based on material currently forthcoming in substantially the same form in *Public Understanding of Science.*
2. The term "effects" implies a one-way relationship between media content and popular thinking that is not supported by contemporary scholarship.
3. This professional ethic in U.S. journalism is generally believed to result from economic necessity rather than moral reasoning. Given the dominance of the Associated Press wire service, the dependence of U.S. news media on advertising revenues for survival, and the fact that few U.S. news markets have more than one daily print outlet, news has become a commodity that must appeal to as broad a range of consumers as possible.
4. Kolata in particular became the subject of considerable controversy in science journalism circles shortly after the peak of the cloning story because of her alleged misuse of scientific sources to support a personal agenda. However, journalists always seek to control both their sources and their stories. Achieving professional prominence is dependent on their doing so successfully, while simultaneously maintaining the appearance of acting merely as neutral observers.

8

The Terminator Gene

It takes millions of dollars and years of research to develop the biotech crops that deliver superior value to growers. . . . These few growers who save and replant patented seed jeopardize the future availability of innovative biotechnology for all growers.

—Monsanto Corporation, 1997 advertisement in *Farm Journal*
(quoted in Crouch 1998)

Whatever the reasons cloning generated such enormous media attention, the controversy seemed to have little lasting impact on either public opinion or the regulatory environment. A *Time*/CNN poll conducted by Yankelovich Partners, Inc., in February 1997 showed that at the height of the cloning furor 66 percent of the American public thought animal cloning was a "bad idea" (for a review of these and the other cloning poll statistics that follow see Singer, Corning, and Lamas 1998). An NBC/*Wall Street Journal* poll conducted by Hart and Teeter Research in March 1997 that used a question that specifically referred to sheep cloning by "a scientist in Scotland" showed that 62 percent thought cloning was a "bad thing;" 50 percent of respondents in a February 1997 ABC poll that used a similarly worded question "disapproved" of animal cloning. However, a January 1991 poll by Princeton Survey Research Associates had shown that 60 percent, nearly the same proportion as the polls contemporary with the controversy, had disapproved of animal cloning *six years earlier.* On the basis of these statistics, it is very hard to credit media accounts with very much influence on public opinion.

Cloning also disappeared from the media agenda almost as quickly as it had appeared and without any new substantial regulations or legislation having been enacted in response. But this does not mean the cloning story was without impact on public thought or public life. It had served to sensitize journalists, editors, and producers—along with their audiences—to the existence of controversy associated with agricultural and pharmaceutical applications of biotechnology. It undoubtedly forged new source-journalist links with academic ethicists, dissenting scientists, and other bona fide authorities representing critical perspectives on the uses of the new genetics, as well as with a broader range of activist sources, making it that much easier for a greater diversity of views to be included in news accounts the next time around and legitimizing the adoption of a critical stance about biotech. And the next time around, in terms of biotechnology-related controversy, was just around the corner.

In the midst of this now increasingly volatile public opinion climate the next agricultural biotechnology dispute soon emerged, the controversy over the terminator technology developed with the cooperation of the U.S. Department of Agriculture, patented by the Delta and Pine Land Company in 1988, and widely associated with Monsanto Corporation. Within a matter of a few months, Monsanto gave every appearance of trying to acquire the patent by making a move to acquire the Delta and Pine Land Company. While the mainstream press still reported primarily a large-institution point of view, dissent was becoming more visible, especially when a broader range of publications is considered. The emerging popularity of the Internet as a channel of communication, especially of minority views, may or may not have changed public opinion in important ways, but it certainly made the observation that U.S. complacency was hardly universal easier to make.

Whatever the reasons, the terminator controversy seemed to mark the "coming of age" of agricultural biotechnology as an area of public debate in which the existence of alternative views could no longer be ignored. A September 1999 Gallup food safety poll that included some questions on biotechnology showed that while a slim minority (51 percent) of Americans now support the use of biotechnology in agriculture, 41 percent of survey respondents are opposed (Saad 1999). Further, strong opponents outnumber strong supporters nearly two to one (16 versus 9 percent). While the proportions, which are always responsive to the particular question wording employed, are not particularly comparable to those in the earlier survey results reported in chapter 4, the same kind of split in U.S. opinion is apparent. More strikingly in the 1999 survey, however, over two-thirds, or 68 percent, would be willing to pay extra for food labeling that indicated foods produced using biotechnology, even though only 27 percent thought that biotech actually represented a "serious health hazard to consumers." This result is difficult to explain except in terms of the controversy's having become legitimized to the point of broad public recognition. Finally, most (82 percent) of the Gallup respondents had heard at least something about food biotechnology, even though only 10 percent reported having heard "a great deal."[1] This also strongly suggests continued media attention to the controversy, although it is not very much higher than the three-fourths of the population who reported having heard something about genetic engineering more generally circa 1987 (chapter 4). It seems reasonable to assert that the cloning controversy may have had a sensitizing effect that spilled over to other applications of biotechnology in agriculture.

As has already been argued here, the widespread assumption that the U.S. public has ever been enormously more willing to accept biotech than the European public is hard to support. However, the existence of controversy in the United States, once absent from the pages of the mainstream papers, is becoming suddenly more conspicuous. This may be, in large part, a media phenomenon rather than an actual public opinion shift, and the cloning controversy—which was among the most prominent U.S. science news stories ever reported—probably made a major contribution to that increased visibility. But making alternative points of view apparent has the potential to counteract spiral of silence effects. Thus the appearance is one of a fairly rapid shift in public opinion, leading to increased public debate that may in turn change actual opinions. Following on the heels of cloning, the terminator controversy followed a different course. The sequence of events here helps illustrate how the agenda-building dynamics that can, over time, disrupt a spiral of silence are likely to be especially sensitive to these kinds of synergistic (rather than isolated) influences. This chapter reviews the ways in which the terminator controversy emerged and was variously reported.

TECHNOLOGY AND CORPORATE STRATEGY

Applying the label "terminator technology" to Delta and Pine Land Company's patent for a way to engineer crops so that they would not produce viable next-generation seeds was apparently the brainchild of the Rural Advancement Foundation International (RAFI), a group that has been highly critical of the projected economic impact of biotechnology on farmers. This technology has been represented as of special concern for the smaller, more marginal, and subsistence farms in the developing world. Although arguably not a great departure from the status quo for corn, in the United States at least a crop already typically planted in hybrid form and therefore not the best candidate for a farmer to replant by saving the seed from a previous year's crop,[2] terminator technology could extend this kind of built-in plant variety protection to other crops such as wheat, rice, soybeans, and cotton (Crouch 1998). While use of this technology could limit the viability of second-generation genetically engineered "volunteer" plants that might otherwise grow from seed that "escapes" from a cultivated field into other areas, thereby possibly dampening some fears that these new varieties could have unpredictable ecological effects, the possibility that terminator plants could cross-pollinate with non-terminator varieties raises new concerns.[3] The outcome of such a cross would not be entirely predictable; the viability of neighboring farmers' plots might be limited, for example, and there could be unforeseen ecological consequences outside of cultivated areas. The main reason for the development of the technology appears to have been to achieve the objective Monsanto described in the ad copy that introduces this chapter: the protection of their own business investment in genetically engineered crop varieties, not the protection of the environment from genetically altered plants. Whether or not this was the case, the perception that Monsanto had nothing but its own self-interests at heart was certainly widespread in media accounts.

With terminator technology, farmers must purchase all seed every year. Without it, they have considerable more independence, perhaps to save and replant one year but purchase seed another year depending on any number of conditions, despite licensing agreements and other kinds of controls that are supposed to prevent the planting of second-generation patented seed whether or not it carries the terminator gene. Alongside fears of ecological and crop damage are fears of a perceived potential for monopolistic domination of United States and world agriculture by a very few seed companies. Whether or not terminator plants have the potential to harm non-terminator neighbors, farmers who do adopt them will clearly have no choice but to repurchase seed the following year. And economic factors could force farmers to adopt this innovation whether they want to or not. These fears may or may not be exaggerated; whether they are or not, these issues have probably contributed to a backlash reaction against genetic engineering of crops more generally. Farmers, consumers, and the public do not always seem to make clear distinctions among various forms of biotech, even though polls have consistently shown differences in attitudes toward various applications.

Just as cloning upset cultural ideas about the biological order, something about terminator technology seems to run against the grain of U.S. culture. Monsanto did not help its public relations effort very much by suing a Canadian farmer for allegedly growing their herbicide-tolerant canola without having purchased the seed; the controversy soon reached the national press, appearing in a February 1999 *Washington Post* article by science writer Rick Weiss that also discusses the "terminator" prospects (Weiss 1999a). The farmer in question, a former mayor who was not about to take Monsanto's attack lying down, claimed that any engineered canola must have arrived on his farm due to wind-borne drift or some other accidental means. According to reports, Monsanto also broadcast radio advertisements in some areas encouraging farmers to expose neighbors who might be planting unlicensed Monsanto seed, even providing a toll-free "tip line" for this purpose. A move more unlikely to appeal to farming communities, which often remain small and relatively close-knit to this day even in the United States, is difficult to imagine.

Despite the fact that most commercially successful farms in the United States are now big businesses, even where they remain family-owned ones, the David-and-Goliath image of a gigantic multinational agribusiness corporation sending out detectives and lawyers to pursue individual farmers accused of planting "illicit" seed on their own farms was not lost in the journalistic accounts that first brought the terminator controversy to the attention of the general public. Indeed, the fact that this agriculture-related controversy, highly technical in both its legal and scientific details, reached the national news at all is remarkable and certainly reflects the way that the story struck a strong chord.[4]

Monsanto's stake in genetic engineering of crops, particularly its Roundup Ready varieties, is complex. Roundup is a popular herbicide on which Monsanto has held the patent; Roundup Ready crops are genetically engineered to withstand heavy application of Roundup and survive.

Monsanto's U.S. patents on the active ingredient in Roundup, glyphosate, will expire in the year 2000 (Monsanto's strategy 1999). In fact, the company has been reported to have signed agreements with several of its heretofore competitors to distribute glyphosate-based herbicides independently. In addition, according to a 1998 Monsanto press release (Monsanto Corporation 1998), the company actually reduced the price of its own Roundup herbicide on September 1 of that year by up to $10 per gallon at the same time it increased its "technology fee" for the herbicide-resistant Roundup Ready soybeans.

In other words, they might be able to make up for the expiration of their glyphosate patent if a reduction in the market price of glyphosate products (which they will soon no longer control) makes them more attractive to farmers, increasing demand for the patented plant varieties that can best withstand their application (which they still do control, in part through their projected ownership of the company that patented the biological mechanism of the so-called terminator gene). Monsanto predicts net savings for the typical grower of from $1 to $3 per acre through the use of the reduced-price Roundup in conjunction with Roundup Ready soybeans, although their posting of this release on the corporate Web site at <http://www.monsanto.com> ends with a note warning that the release "contains certain forward-looking statements" and explains that "actual results may differ materially from those anticipated." Nevertheless, taking these claims at face value, Monsanto seems certain enough of the profits its new crop varieties promise to bring in to, in effect, pay farmers to try them. If they are successful, part of this cost will be borne by the other companies who will be selling Roundup-like products at prices depressed by the rate cut, whereas all of the profits generated by the new patented seed will belong to Monsanto. This becomes a substantially more risky venture if the cheaper Roundup can be applied with equal freedom to soybeans grown with seed Monsanto has not licensed, which is what the terminator gene could have helped to guard against.

Concerns about the economic impact of the terminator technology Monsanto may have hoped would protect its strategy are not confined to a handful of radicals. In June 1999, the enormously well-known and widely respected Rockefeller Foundation appealed to Monsanto to halt deployment of the terminator, citing worry over the damage being done to public support for crop biotechnology in general (Kilman 1999). This was an unprecedented type of appeal. In response, in an October 4 open letter from Monsanto CEO Robert B. Shapiro to Rockefeller Foundation president Gordon Conway, Monsanto made clear its "public commitment not to commercialize gene protection systems that render seed sterile," an implicit capitulation not so much to the fact this technology threatened poor farmers in less developed parts of the world, but to the much more self-evident claim that the company's involvement with the development of the technology was in danger of becoming a public relations nightmare. The corporation most closely associated with agricultural biotechnology, and one which had founded its public relations battles up until then largely on one-sided messages about biotech's benefits, was forced into retreat in the face of widespread public criticism. It hardly mattered to the outcome how much substance those criticisms did or did not contain.

BREAKING INTO PRINT

On February 8, just five days after Weiss's *Post* story on Monsanto's "gene police" tactics, the paper published a second article by Weiss, this time specifically on the debate about terminator technology (1999b, A9). Whether and in what form the terminator debate would otherwise have come to light, it is difficult to avoid the connection between the news attention given to Monsanto's enforcement strategies and to the longer-term prospects for genetic control through terminator technology. Not all the news coverage painted Monsanto as a corporate power throwing its weight around. Weiss's original article includes several quotes from Monsanto spokespeople discussing the benefits and popularity of its herbicide-resistant seeds. A February 1 article in *Time* entitled "The Suicide Seeds" starts out describing Monsanto as "the best place to turn for help these days" for farmers who want good crops (Kluger 1999). A February 4 article in the *Lincoln Journal Star* likens saving patented seed to "stealing an unpublished novel" by photocopying it (Hovey 1999). Nevertheless, the amount of attention given to the terminator itself implies the existence of controversy.

This February news blitz was not the first time the terminator was publicly discussed. For example, nearly a year earlier, on May 25, 1998, the *Wichita Falls Times Record News* printed a story datelined Haskell, Texas, on the development by Delta and Pine Land and the U.S. Department of Agriculture (Noack 1998). The story mentions Monsanto's interest in acquiring Delta and Pine Land and one farmer's associated concern about monopolization in the seed business, but only in passing in a single paragraph some distance into the piece. (The government scientist who participated in the terminator development was based in Lubbock, near North Texas cotton country.) The Raleigh, North Carolina, *News & Observer* covered it in November, shortly after a vote to condemn it by the Consultative Group on International Agricultural Research meeting at the World Bank (Williams 1998a, 1998b), as did the *St. Louis Post-Dispatch*, hometown paper for Monsanto's own headquarters (Lambrecht 1998). But the story did not appear to become big in the national news until the first week of February 1999. By June of that year, the Rockefeller Foundation was calling for Monsanto to call a halt to its pursuit of terminator commercialization, and by October, the company had capitulated.

What caused the terminator to finally become national news? Like many news stories, those appearing in February are difficult to connect to a specific event. The Canadian canola farmer was neither the first nor the last to be subject to intense Monsanto scrutiny, although he was probably one of the more difficult to intimidate. The *Time* magazine piece—by date, the first of the February contingent—is not about him. But as its author suggests, to understand "the p.r. beating Monsanto is taking," we need to "check out the Web." Beneath the nearly undisturbed calm in the national news seethed a cauldron of criticism and concern much more visible on the Internet and in alternative and specialist publications, a cauldron that had probably been boiling beneath the surface of mainstream news accounts for some time.

ALTERNATIVE ACCOUNTS

The February 1 *Time* piece gives 4,000 as the number of people who had responded electronically to RAFI's call for terminator opponents to flood the Department of Agriculture with protest letters, perhaps unimpressive in comparison to the 140,000 later reported to have expressed their opinions about proposed rules allowing (among other things) biotechnology-based foods to be labeled organic just a few months later in May (Manning 1998), but apparently enough to grab *Time*'s attention. In addition to RAFI, the Internet has given a powerful voice to activist groups with names like Mothers for Natural Law and Ethical Investing, as well as the more familiar players such as Greenpeace and the Union of Concerned Scientists. The Net allowed small businesses such as Vermont's Gardener's Supply Company to enter the fray in new ways; on January 8, 1999, they had posted a release entitled "There's No Fooling Mother Nature! Say NO to the Terminator Gene" on the Web alongside their supply catalogue for environmentally aware home gardeners (www.gardeners.com/prtermgene.html). Another company called Garden City Seeds offered $50 packages of 26 "good old fashioned standard varieties" of seed, designed to be enough to grow food for a family of four for a year, in response to the terminator crisis. Net-based gardeners' groups buzzed with discussions highlighting differences between hybrids, traditional open-pollinated "heirloom" varieties, and terminator plants. Press releases from India and the rest of the developing world, often denied space in the wire-dominated, mainstream U.S. press system, are just as accessible on the net as those from within U.S. borders and from the headquarters of U.S. corporations. And these sources have been consistently critical of the potential impact of the terminator on small farmers.

Ironically, the Internet technology some have feared would facilitate corporate domination certainly appears here to have given voice to the grassroots terminator opposition that eventually broke through into the mainstream news. But the emerging criticism was not confined to this new medium. The alternative press in the United States, as weak and nearly invisible as it seems to have become, also became a player. *Mother Jones* has run a number of articles concerned about the whole area of genetic engineering, particularly about the organic food controversy. Less well-recognized alternative voices ranging from *Earth Island Journal* to *Global Pesticide Campaigner* have also run pieces about the terminator.

The British magazine *The Ecologist* published an entire issue in September/October 1998 about alleged Monsanto malfeasance; the issue became somewhat of an underground cause celebre for the entire first printing having been destroyed, allegedly out of fear of the corporation's wrath. (British libel law is somewhat less protective of its press in such cases than that in the United States.) One of the offending pieces was entitled "Terminator Technology: The Threat to World Food Security" (Steinbrecher and Mooney 1998). Its authors argue that hybridization, to which terminator technology is often compared, had also been a "biological weapon" designed to prevent farmers from saving and replanting their own seed. An article in *The Nation* called a series of corporate ads that appeared in British papers and were designed to respond to such assertions "greenwashing"; according to the article, the ads included contact information for anti-biotechnology groups like Friends of the Earth and Greenpeace (Margaronis 1998). Apparently, the world of corporate biotechnology had finally discovered the principle of using two-sided messages to persuade, but it appeared to be a case of too little, too late, and the campaign's implied alignment with the environmentalist world too transparent to be effective.

The trade press was largely silent on the whole constellation of issues associated with plant patent protection, including the terminator. For example, *Chemical & Engineering News* reported the Monsanto suit of Canadian farmer Percy Schmeiser, but apparently not until May 1999, three months after its appearance in *Time* and the *Washington Post* (Monsanto in dispute 1999). *Progressive Farmer*'s March 1999 piece on "Monsanto's Strategy for Roundup" manages not to discuss what role the company had expected terminator technology to play (if there was one) at all. The U.K.-published *Outlook on Agriculture,* however, which quite regularly seems to print articles about the biotechnology successes claimed by companies like Monsanto and Novartis, included the concerns of RAFI about terminator technology and the Indian government's ban of it in a piece entitled "Terminator Tribulations" (1999), lending the impression that a wider range of perspectives is available in its pages than may be visible in equivalent U.S. trade publications.

RAFI repeatedly emerges as a key source of anti-terminator views, but it is hardly acting either alone or in a vacuum. Activist messages helped mainstream skepticism about the terminator emerge, but they did so in the wake of the cloning controversy, which had created an environment of increased concern about biotechnology generally. Internet-based communications seemed to facilitate the spread of alternative and non-Western points of view in an era in which the activist press in the United States seems to have grown weak. Concerns that this technology has been colonized primarily by advertisers and marketers for their own purposes are well supported by the proliferation of public relations-oriented corporate Web sites, as well as sites specifically designed to sell products. But there can also be little doubt that Internet technology is filling part of the gap between the range of the opinions the U.S. press accommodates and the range that actually exists, and—in the case of the terminator—Internet discourse may well have helped frame the agenda of the mainstream press. As more and more journalists turn to the Internet for information that is easily accessible and cheap, the homogeneity of U.S. news may be diminishing.

Corporate (and researcher) messages that attempted to sell the terminator as environmentally sound because it could prevent genetically modified seed from escaping to other areas did not appear to be effective in deflecting criticism; neither, apparently, were attempts to acknowledge the controversy and associate Monsanto with environmentalist and sustainable agriculture interests in the light of repeated reports of Indian farmers burning Monsanto crops. The terminator never made a very big ripple in the mainstream news. But rumors continue to surface that biotechnology-based crops were not doing as well as expected, and farmers have begun to express reservations about the prospects that markets for these crops might become increasingly constricted due to public pressure. Monsanto's responses may or may not have been adequate to dampen these fears. By December 1999, news reports were discussing whether the company would decide to break itself up. Its stock had fallen. While many in the scientific community continue to hold high hopes for agricultural biotechnology's ability to improve human health and nutrition, one of its primary business champions does not appear to be doing as well as it had hoped. If new media technology and the Internet helped this and other companies to spread their corporate messages, they have also facilitated the spread of activist critiques.

PROSPECTS AND PERCEPTIONS

Whatever the reality, in the late twentieth century U.S. agricultural biotechnology in the guise of the terminator became perceived as a tool of large-scale (if not monopolistic) capitalism rather than of humanitarian progress. Protests by subsistence farmers on the other side of the globe and by organic home vegetable gardeners here in the United States both played a role in this outcome, aided by communications technology (the Internet and the World Wide Web) that seems to be filling the void created by a news system grown increasingly homogeneous and monolithic. But these events occurred in a climate of public opinion in which public thinking about corporate involvement with biotechnology had been framed by the events surrounding bovine somatotropin, and public thinking about biotechnology itself had been more recently reshaped by the reporting of events surrounding cloning.

While it would be greatly overstating the case to attribute these outcomes to communication patterns primarily, it is likely mass communication that created the spiral of silence that once existed in the United States regarding agricultural applications of the new genetics. It may be accurate to state that the U.S. population has not been concerned about genetically modified food issues to the same degree as the populations in much of Europe, but it has never been accurate to state that the U.S. population has not contained significant numbers of people who either were concerned or were subject to becoming concerned once information about genetically modified foods and other agricultural biotechnology reached them. Religious traditionalists in the United States were unlikely to approve of science playing God; the less traditional but environmentally conscious segment were unlikely to approve of corporate scientists tinkering with ecological systems and processes. Both groups hold biology to be somewhat sacred and large institutions somewhat suspect. The emergence of objections seems, in retrospect, to have been inevitable. Recent reports of bioengineered pest-resistant crops damaging fragile monarch butterfly populations seem like the perfect symbolic representation of the general opinion atmosphere—an atmosphere of suspicion.

Media processes surrounding how biotechnology was publicly represented were not without effect, although these effects were not what are typically attributed to the media. Their impact was apparent primarily in reverse; they constructed a picture of a public tolerant of genetically modified foods, for example, if not enthusiastic about them, and a world in which criticism and concern was confined to a handful of extremists. The perception that the U.S. public was generally complacent about agricultural biotechnology, based in part on media mainstreaming and its creation in turn of a spiral of silence effect, probably helped fuel the current backlash evident in the European and Asian reactions, as well as the now much more visible pockets of concern within the United States, pockets that are likely to continue their growth. It is possible to speculate that had corporate complacency not led to the attribution of objections to bovine somatotropin and other early food-related biotechnology to ignorance, and to the use of one-sided messages about biotechnology's promise that were designed to combat this ignorance rather than contribute to meaningful public debate, the outcome could have been different. It is also possible to speculate, however, that had Ian Wilmut's cloning technology not unleashed an almost unprecedented flurry of news attention, the extent to which genetic engineering challenges some of our most cherished values—including human individuality and personal autonomy—might have remained unarticulated for some time yet. And it is also possible to imagine that had the Internet not played a role in articulating and spreading a wider spectrum of views than was available in the mainstream press, this debate would have remained a private one for much longer. In the long run, this would probably have been no better from the point of view of those who believe biotechnology is our future. It might actually have been considerably worse.

NOTES

The information included in this chapter about how various news outlets reported the terminator gene controversy were gathered and analyzed with the assistance of students in an undergraduate media research methods course taught at Texas A&M University in the summer of 1999: Kellee Ashby, DaLynn Barker, Samara Callaway, Amy Conrey, Beverly Cooper, Josh Dixon, Truman Glenn, Jennifer Hancock, Shanda Lynn, Josh Mills, Faith Patton, Michelle Peltier, Jennifer Piazza, Naida Robles, Brooke Saucier, Sundie Savage, Brooke Sullivan, Jill Taylor, Sam Willson, and John Wyatt.

1. The Gallup report also states that opposition to food biotechnology was strongest among lower-income and less-educated Americans and among those who reported having heard the least about the issues, points that would certainly appear to support the attribution of opposition to inadequate or inaccurate information. This is not the general perspective adopted in this book; while some data supporting either view is available, other evidence (as presented in earlier chapters) finds more opposition among the better-informed, although their views are not necessarily based on health or safety concerns. However, the appropriate interpretation of the Gallup data, based on questions asked in the context of the general topic of food safety, is unclear. These patterns may emerge differently in the context of a food safety discussion than when asked in another context, and the influence of a rapidly changing climate of public opinion on people hearing of a controversy for the first time is unknown.

2. Hybrid seed produces plants with predictable qualities; seed from plants grown with the hybrid seed is less predictable. Seed from plants with the terminator gene are intended to be incapable of growing at all.

3. Proponents of herbicide use—usually its marketers—also argue the practice permits reduced cultivation for weed control and thus could help conserve soil.

4. According to news accounts, the farmer—68-year-old Percy Schmeiser of Bruno, Saskatchewan—is defending himself. A Canadian judge is scheduled to rule on the case later this year (Walker 2000).

9

Lessons and Directions

American values such as individuality and personal autonomy are alive and well. Our respect for genetic identity and for ecological integrity may be two halves of the same coin. Underneath our apparent penchant for environmental destruction may lie more reverence for biology than we had quite realized. Science does not "sell" very easily if it is seen as conflicting with these kinds of fundamental beliefs. On the other hand, the cloning furor demonstrated that vigorous public debate does not necessarily turn public opinion permanently sour. Nevertheless, genetically modified foods, which originally seemed unlikely to engender much debate at all in this country in comparison to Europe and elsewhere in the world, now appear likely to meet with a much rockier fate. It is tempting to speculate about what the current climate of public opinion in the United States would have been like had spiral of silence effects, aided by a news system that overrepresents large-institution points of view, not made it appear there was less opposition than there is, effectively suppressing public debate. In fact, we are different in our thinking only possibly in degree, but not so clearly in kind, as the populations of many European nations. Opinion here is, and has always been, divided. The agenda-building process for these issues has been slow, but it rests on underlying public concerns that have been detectable for quite some time. Media messages did not create these concerns, which are rooted in deeply held values and perceptions unlikely to be easily changed.

Biotechnology in U.S. agriculture has been promoted using one-sided messages that assumed most opposition surrounded health and safety issues, when in fact socioeconomic and environmental-ecological issues explain many reservations, along with perceptions of ethics, openness, and regulatory adequacy. More recent attempts to associate biotech with responsible environmental stewardship and feeding the world's hungry, whether the technology really has that potential or not, do not appear to have been particularly effective. It is nearly impossible to even guess whether these strategies would have had more impact earlier, but if they were delivered as one-way messages about benefits they might not have helped biotech's proponents at all in the longer term. Observers might take a certain comfort in seeing that the American public is apparently not so easy to fool as we might have thought. To persuade them of biotech's benefits, it is likely that it would have been necessary to demonstrate a real willingness to accommodate their concerns through product labeling and other more open forms of communication, as well as by demonstrating an authentic willingness to address the full range of issues being raised. It may or may not be too late to pursue this path.

Proponents of increased public involvement in science policy-making come from both the right and the left of the political spectrum. No one has a formula for making this work. Activist involvement, even legitimized activist involvement, is not quite the same thing. But the available evidence to date suggests that only through engaging the public in active debate is something as fundamentally novel as genetic engineering likely to be accepted. Ironically, the European experience—which many in the United States have perceived to be an irrational and anti-scientific response—may be preferable from the point of view of biotechnology's proponents, and at the same time as it is preferable from the perspective of promoting democratic practice. This is not to say that the Europeans will eventually accept genetically modified foods; they may not. But the kind of paternalism that distrusts public debate on issues involving science and technology policy and seeks instead to manipulate public opinion to some particular foregone conclusion is unlikely to be successful, long term, in a modern democracy. Public debate is a healthy thing for issues involving science and technology, as for other kinds of issues.

A significant unresolved question is that of the relationship between scientific education, or more specifically, knowledge of biotechnology, and acceptance. Highly educated scientists do not agree on these issues, suggesting that dissent cannot be educated away. Many, perhaps most, significant disputes about the use of science and technology in society center on issues of ethics, equity, and justice, and how to choose the most prudent collective course of action in the face of great uncertainty, not necessarily on the science itself. Nevertheless, some studies have suggested a relationship between educational level or background scientific knowledge and opinions in this area. It remains quite possible that some concerns—such as whether genetically modified foods are safe to eat—might be most likely to arise among those who have most recently heard about the idea, or who have less scientific background, while others—such as the more subtle dangers involving the risks of monoculture, loss of crop genetic diversity, and complex ecological impacts—are of greater concern to those who have more knowledge of the relevant science, rather than less. This is not to even consider how opinions about ethical or economic dimensions might behave. Complicating this picture further is the fact that science itself is not neutral in this instance; the policy positions of scientists seem to depend in part on the degree to which they have a personal stake in biotechnology-related research. So if ever there is a time when science can be called on to resolve political disputes—and this is quite doubtful—that time is not now, and not with respect to biotech.

The relationship between science literacy and opinions about science policy needs to be further explored, and this exploration will involve a close reexamination of what we mean by science literacy in the first place. Scientists often seem to be concerned that nonscientists will come up with wrong opinions on irrational grounds, leading to the conclusion that the public is best not consulted on matters scientific and technical. It is probably true that there is a lot of ignorance in the world still. But ethical and political philosophy aside, excluding ordinary people from science policy debate is no longer a practical strategy. The public is simply not going to put up with it, and the risk of a backlash is too great. Biotechnology illustrates this reality particularly well. We need to engage, not exclude, the public in this kind of debate or risk the consequences, which are no less than eventual rejection of much new science and technology. Of course, it is better if the public understands the science itself as well as possible. But these debates cannot be made to disappear through education, and they cannot and should not be limited to expert voices.

What do we mean by science literacy in the first place? Available measures generally look at access to scientific facts, and to some extent familiarity with principles such as the scientific method. But the ability to observe, test, question, learn, and reason is not reducible to having access to a certain store of facts, and the ability to make good use of information and participate wisely in decision-making cannot in turn be reduced to the ability to observe, test, question, learn, or reason. In both making public policy and designing good education, we need to think more deeply about what it means to be wise. Wisdom is something likely best learned through experience, but if science education can be broadened to encompass a healthy respect for wisdom and if science communication can recognize its relevance, as a society we will be better off.

From a communications perspective, the wisest thing we can do is to encourage the broadest possible public debate. And this is true whether we hope to promote biotechnology or to foster democracy. News that presents science always as a fait accompli and that presents only one perspective on its best use is not very helpful in achieving either of these goals. The present U.S. news system may not be very well adapted to presenting it in any other way, and if so, new channels must be developed. This is not to suggest that we can get the American public to accept new technologies simply by nurturing the illusion that they have a choice. But without their having a choice, we have failed on both counts; we are unlikely to win their support for science and technology, and we are certainly not making democracy work any better. We do need a scientifically literate public, but we also need to reexamine what we mean by that phrase. The sagacity to know how to use science well is not so easily memorized as the periodic table of elements or the formula for the conversion of mass to energy.

References

Alger, Dean. 1998. *Megamedia: How Giant Corporations Dominate Mass Media, Distort Competition, and Endanger Democracy*. Lanham, Maryland: Rowman & Littlefield.

American Corn Growers Association. Undated. *GMO Brochure,* Section 6. <http://www.acga.org/GMOBrochure/06.htm>.

Bagdikian, Ben. 1983. *The Media Monopoly*. Boston: Beacon Press.

Beck, Ulrich. 1992. *Risk Society: Towards a New Modernity*. Newbury Park, CA: Sage.

Bitter Fruit. 1997. *Wall Street Journal*, 4 February.

Boulter, D. 1995. Plant Biotechnology: Facts and Public Perception. *Phytochemistry* 49(1): 1–9.

Bradbury, Judith. 1989. The Policy Implications of Differing Concepts of Risk. *Science, Technology, & Human Values* 14(4): 380–399.

Brookes, Rod. 2000. Tabloidization, Media Panics, and Mad Cow Disease. Pages 195–209 in Colin Sparks and John Tulloch, eds. *Tabloid Tales: Global Debates over Media Standards*. Lanham, MD: Rowman & Littlefield Publishers.

Carson, Rachel. 1962. *Silent Spring*. Boston: Houghton Mifflin.

Chemical & Engineering News. 1987. Public's Feelings about Biotech Are Mixed. 1 June: 22.

Chemical Week. 1987. Poll Results May Bode Well for Biotechnology. 10 June: 15–16.

Clemens, Elisabeth S. 1986. Of Asteroids and Dinosaurs: The Role of the Press in the Shaping of Scientific Debate. *Social Studies of Science* 16(3): 421–56.

Cloning for Good or Evil. 1996. *New York Times*, 25 February.

Crouch, Martha L. 1998. How the Terminator Terminates: An Explanation for the Non-Scientist of a Remarkable Patent for Killing Second Generation Seeds of Crop Plants. Occasional paper of The Edmonds Institute, 20319 92nd Avenue West, Edmonds, Washington 98020.

Davison, Aiden, Ian Barns, and Renato Schibeci. 1997. Problematic Publics: A Critical Review of Surveys of Public Attitudes of Biotechnology. *Science, Technology, & Human Values* 22(3): 317–48.

DeFleur, Melvin L., and Sandra Ball-Rokeach. 1989. *Theories of Mass Communication*. 5th ed. New York: Longman.

Dickson, David. 1984. *The New Politics of Science*. New York: Pantheon.

Durant, John, Martin W. Bauer, and George Gaskell, eds. 1998. *Biotechnology in the Public Sphere: A European Sourcebook*. London: Science Museum.

Edelman, Murray J. 1971. *Politics as Symbolic Action: Mass Arousal and Quiescence*. Chicago: Markham.

Eden, Michael J. and John T. Parry. 1996. *Land Degradation in the Tropics: Environmental and Policy issues*. London: Pinter.

Ellul, Jacques. 1965. *Propaganda: The Formation of Men's Attitudes*. New York: Knopf.

Epstein, Samuel. 1999. Monsanto's Genetically Modified Milk Ruled Unsafe by the United Nations. *PRN Newswire*, 18 August <http://www.preventcancer.com/PRS/august 24.html>.

Evans, Geoffrey, and John Durant. 1995. The Relationship between Knowledge and Attitudes in the Public Understanding of Science in Britain. *Public Understanding of Science* 4: 57–74.

Evans, William. 1995. The Mundane and the Arcane: Prestige Media Coverage of Social and Natural Science. *Journalism & Mass Communication Quarterly* 72(1) (spring): 168–77.

FDA Warns the Dairy Industry Not to Label Milk Hormone-Free. 1994. *New York Times*, 8 February.

Frewer, L. J., D. Hedderley, C. Howard, & R. Shepherd. 1997. "Objection" Mapping in Determining Group and Individual Concerns Regarding Genetic Engineering. *Agriculture and Human Values* 14: 67–79.

Frewer, Lynn J., Chaya Howard, and Jackie I. Aaron. 1998. Consumer Acceptance of Transgenic Crops. *Pesticide Science* 52: 388–393.

Friedman, Sharon M., Sharon Dunwoody, and Carol L. Rogers, eds. 1999. *Communicating Uncertainty: Media Coverage of New and Controversial Science*. Mahwah, N. J.: Erlbaum.

Gandy, Oscar H. 1982. *Beyond Agenda-Setting: Information Subsidies and Public Policy*. Norwood, New Jersey: Ablex.

Gans, Herbert J. 1980. *Deciding What's News: A Study of CBS Evening News, NBC Nightly News, Newsweek, and Time*. London: Constable.

Gaskell, George, Martin W. Bauer, and John Durant. 1998. Public Perceptions of Biotechnology in 1996. In *Biotechnology in the Public Sphere: A European Sourcebook*, ed. John Durant, Martin W. Bauer, and George Gaskell, 189–214. London: Science Museum.

Gaskell, George, Martin W. Bauer, John Durant, and Nicholas C. Allum. 1999. Worlds Apart? The Reception of Genetically Modified Foods in Europe and the U.S. *Science* 285(5426): 384–87 (16 July 1999).

Germany Gets in a Lather about Altering Beer Genes. 1996. *Wall Street Journal*, 8 July.

Gitlin, Todd. 1980. *The Whole World is Watching: Mass Media in the Making and Unmaking of the New Left*. Berkeley: University of California Press.

Glickman, Dan. 1999. *New Crops, New Century, New Challenges: How Will Scientists, Farmers, and Consumers Learn to Love Biotechnology and What Happens if They Don't?* <http://www.usda.gov/news/releases/1999/07/0285>.

Goodell, Rae. 1977. *The Visible Scientists*. Boston: Little, Brown.

Grunig, James E. 1989. Symmetrical Presuppositions as a Framework for Public Relations Theory. In *Public Relations Theory*, ed. Carl H. Botan and Vincent Hazleton, Jr. Hillsdale, New Jersey: Erlbaum.

Gutteling, Jan M., Anna Olofsson, Bjorn Fjaestad, Matthias Kohring, Alexander Goerke, Martin Bauer, and Timo Rusanen. 1999. Development of the Coverage of Modern Biotechnology in the Opinion-Leading European press 1973–1996. Unpublished paper. University of Twente, Department of Communication Studies, PO Box 217, 7500 AE Enschede, the Netherlands.

Hachten, William A., and Harva Hachten. 1999. *The World News Prism: Changing Media of International Communication*. 5th ed. Ames: Iowa State University Press.

Hillers, Virginia N., and Michiel R. H. Löwik. 1998. Incorporation of Consumer Interests in Regulation of Novel Foods Produced with Biotechnology: What Can Be Learned from the Netherlands? *Society for Nutrition Education* 30(1) (January–February): 2–7.

Hoban, Thomas J., and Lisa D. Katic. 1998. American Consumer Views on Biotechnology. *Cereal Foods World* 43(1): 20–22.

Hornig, Susanna. 1990. Science Stories: Risk, Power and Perceived Emphasis. *Journalism Quarterly* 67(4)(winter): 767–76.

Hornig, Susanna. 1992a. Framing Risk: Audience and Reader Factors. *Journalism Quarterly* 69(3)(autumn): 679–89.

Hornig, Susanna. 1992b. Gender Differences in Responses to News about Science and Technology. *Science, Technology, & Human Values* 17(4)(autumn): 532–42.

Hornig, Susanna. 1993. Reading Risk: Public Response to Print Media Accounts of Technological Risk. *Public Understanding of Science* 2: 95–109.

Hotz, Robert Lee. 1997. With Cloning Success, Ethics Issues Intensify. *Los Angeles Times*, 24 February.

Hovey, Art. 1999. Crop Protection. *Lincoln Journal Star* (Nebraska), 4 February.

Hubbard, Ruth. 1995. *Profitable Promises: Essays on Women, Science, and Health*. Monroe, Maine: Common Courage Press.

Hubbard, Ruth, and Elijah Wald. 1993. *Exploding the Gene Myth: How Genetic Information is Produced and Manipulated by Scientists, Physicians, Employers, Insurance Companies, Educators, and Law Enforcers*. Boston: Beacon Press.

Iyengar, S., and D. R. Kinder. 1987. *News That Matters: Television and American Opinion*. Chicago: University of Chicago Press.

Kilman, Scott. 1999. Foundation Asks Monsanto to Kill 'Terminator Gene.' *The Wall Street Journal*, 28 June.

Kluger, Jeffrey. 1999. The Suicide Seeds. *Time* 153(4), 1 February.

Kolata, Gina. 1997. Scientist Reports First Cloning Ever of Adult Mammal. *New York Times*, 23 February.

Lacy, William B., Lawrence Busch, and Laura R. Lacy. 1991. Public Perceptions of Agricultural Biotechnology. In *Agricultural Biotechnology: Issues and Choices*, ed. Bill R. Baumgardt and Marshall A. Martin, 139–62. West Lafayette, Indiana: Purdue University Agricultural Experiment Station.

Lambrecht, Bill. 1998. Critics Vilify New Seed Technology that Monsanto May Soon Control. *St. Louis Post-Dispatch*, 1 November.

Lang, Gladys E., and Kurt Lang. 1983. *The Battle for Public Opinion: The President, the Press, and the Polls during Watergate*. New York: Columbia University Press.

Langreth, Robert. 1997. Cloning Has Fascinating, Disturbing Potential. *Wall Street Journal*, 24 February.

Lasswell, Harold. 1949. The Structure and Function of Communication in Society. In *Mass Communication*, ed. William Schram. Urbana: University of Illinois Press.

Leahy, Peter J., and Allan Mazur. 1980. The Rise and Fall of Public Opposition in Specific Social Movements. *Social Studies of Science 10 (3):* 259–84.

Lewenstein, Bruce V. 1995. From Fax to Facts: Communication in the Cold Fusion Saga. *Social Studies of Science* 25: 403–36.

Lewenstein, Bruce, Tracy Allaman, and Shobita Parthasarathy. 1998. Historical Survey of Media Coverage of Biotechnology in the United States, 1970 to 1996. Paper presented at annual meeting, Association for Education in Journalism and Mass Communication, Baltimore, Maryland, August 1998.

Manning, Anita. 1998. Changes in Store for Organic Food: 140,000 Answer USDA Call. *USA Today*, 4 May.

Margaronis, Maria. 1998. Greenwashed. *The Nation*, 19 October: 10.

Maugh, Thomas H. II. 1997. Scientists Report Cloning Adult Mammal. *Los Angeles Times*, 23 February.

McCombs, Maxwell. E. 1992. Explorers and Surveyors: Expanding Strategies for Agenda-Setting Research. *Journalism Quarterly* 69(4)(winter): 813–24.

McCombs, Maxwell E., and Donald L. Shaw. 1972. The Agenda-Setting Function of Mass Media. *Public Opinion Quarterly* 36: 176–87.

Monsanto Corporation. 1998. *Monsanto Reduces Price of U.S. Roundup Herbicide Brands by $6 to $10/Gallon.* 1 September <http://www.monsanto.com/monsanto/mediacenter/98/98sep1_Rrprice.html>.

Monsanto Corporation. 1999. *Recent Information Alleging Safety Concerns by United Nations Is a Fabrication.* <http://www.monsanto.com/monsanto/mediacenter/1999/99aug18_unitednations.html>.

Monsanto in Dispute with Canadian Farmer. 1999. *Chemical & Engineering News*, 10 May.

Monsanto's Strategy for Roundup. 1999. *Progressive Farmer*, March.

National Science Board. 1998. *Science and Engineering Indicators–1998*. Arlington, Virginia: National Science Foundation (NSB 98-1).

Nelkin, Dorothy. 1995. *Selling Science: How the Press Covers Science and Technology.* 2nd ed. New York: W. H. Freeman.

Nelkin, Dorothy, and M. Susan Lindee. 1995. *The DNA Mystique: The Gene as a Cultural Icon*. New York: W. H. Freeman.

Nestle, Marion. 1998. Food Biotechnology: Labeling Will Benefit Industry as Well as Consumers. *Nutrition Today* 33(1)(January–February): 6–12.

Next, Really Prolific Cows. 1997. *Los Angeles Times*, 25 February.

Noack, Hanaba Munn. 1998. Sowing Seeds of Change. *Wichita Falls Times Record News*, May 25.

Noelle-Neumann, Elisabeth. 1993. *The Spiral of Silence: Public Opinion—Our Social Skin*. 2nd ed. Chicago: University of Chicago Press.

O'Keefe, Daniel J. 1990. *Persuasion: Theory and Research*. Newbury Park, California: Sage.

Plein, L. Christopher. 1991. Popularizing Biotechnology: The Influence of Issue Definition. *Science, Technology, & Human Values* 16(4): 474–90.

Priest, Susanna H. 1995. Information Equity, Public Understanding of Science, and the Biotechnology Debate. *Journal of Communication* 45(1): 39–54.

Priest, Susanna H., and Allen W. Gillespie. 1999. Seeds of Discontent: Scientific Opinion, the Mass Media and Public Perceptions of Agricultural Biotechnology. Invited paper presented at International Botanical Congress meeting, St. Louis, Missouri, July.

Priest, Susanna H., and Jeffrey Talbert. 1994. Mass Media and the Ultimate Technological Fix: Newspaper Coverage of Biotechnology. *Southwestern Mass Communication Journal* 10(1): 76–85.

Priest, Susanna, and Karen Taylor. 1995. The Double Helix and the Popular Imagination: Fatalism and Certainty in Public Understanding of Genetics. Discussion Paper CBPE 95-10, Center for Biotechnology Policy and Ethics, Texas A&M University, October.

Roush, W. 1991. Who Decides about Biotech? *Technology Review* (July): 29–36.

Saad, Lyndia. 1999. What Biotech Food Issue?" *Gallup Organization Poll Release for October 5*; <www.gallup.com/poll/releases/pr991005.asp.>

Schiller, Herbert I. 1992. *Mass Communications and American Empire*. 2nd ed. Boulder: Westview Press.

Schudson, Michael. 1978. *Discovering the News: A Social History of American Newspapers*. New York: Basic Books.

Scottish Institute Seeks Patents for Cloning Process. 1997. *Wall Street Journal*, 9 May.

Shiva, Vandana. 1998. Undated statement prepared for delivery at International Conference on Women in Agriculture, June 28–July 2, Washington, D.C. <http://www.oneworld.org/guides/biotech/Vandana.htm>.

Shoemaker, Pamela J. and Stephen D. Reese. 1996. *Mediating the Message: Theories of Influence on Mass Media Content*, 2nd ed. New York: Longman.

Singer, Eleanor, Amy Corning, and Mark Lamas. 1998. The Polls—Trends: Genetic Testing, Engineering, and Therapy. *Public Opinion Quarterly* 62: 633–64.

Slovic, Paul, Baruch Fischoff, and Sarah Lichtenstein. 1979. Rating the Risks. *Environment*, 21(3): 14–20, 36–39.

Sparks, P., R. Shepherd, and L. J. Frewer. 1994. Gene Technology, Food Production, and Public Opinion: A U.K. Study. *Agriculture and Human Values* 11(1): 19–28.

Steinbrecher, Ricarda A., and Pat Roy Mooney. 1998. Terminator Technology: The Threat to World Food Security. *The Ecologist* 28(5)(September/October): 276–78.

Sterckx, Sigrid. 1997. *Biotechnology, Patents, and Morality*. Edited by Sigrid Sterckx. Brookfield, Vermont: Ashgate.

Taylor, Charles Alan, and Celeste Michelle Condit. 1988. Objectivity and Elites: A Creation Science Trial. *Critical Studies in Mass Communication* 5: 293–312.

Terminator Tribulations. 1999. *Outlook on Agriculture* 28(1).

Thompson, Paul B. 1997. Science Policy and Moral Purity: The Case of Animal Biotechnology. *Agriculture and Human Values* 14: 11–27.

Tuchman, Gaye. 1978. *Making News: A Study in the Social Construction of Reality*. New York: Free Press.

U.S. Congress, Office of Technology Assessment. 1987. New Developments in Biotechnology. Background paper: Public Perceptions of Biotechnology. U.S. Department of Commerce, National Technical Information Service, Publication PB87-207544. Washington, D.C.: U.S. Government Printing Office.

U.S. Congress, Office of Technology Assessment. 1992. A New Technological Era for American Agriculture. Washington, D.C.: U.S. Government Printing Office.

Uttig, Peter. 1993. *Trees, People and Power: Social Dimensions of Deforestation and Forest Protection in Central America*. London: Earthscan.

Walker, Ruth. 2000. Seeds of Doubt in Patent Case. *The Christian Science Monitor*, 7 June.

Weiss, Rick. 1999a. Seeds of Discord. *The Washington Post*, 3 February.

Weiss, Rick. 1999b. Sowing Dependency or Uprooting Hunger? *The Washington Post*, 8 February.

Williams, Bob. 1998a. 'Terminator Technology' Could Curtail Brown-Bagging. Raleigh, North Carolina, *News & Observer*, 8 November..

Williams, Bob. 1998b. Seed-Saving Farmers Forced to Alter Their Ways. Raleigh, North Carolina, *News & Observer*, 8 November.

Index

About the Author

Susanna Horning Priest, Ph.D., is an associate professor of journalism at Texas A&M University, former director the university's Center for Science and Technology Policy and Ethics, and associate editor of the journal *Public Understanding of Science.* She has been doing research on public perceptions of technological risk, risk communication, and associated ethical issues since the late 1980s.